Contact Angle, Wettability and Adhesion

Volume 5

Contact Angle, Wettability and Adhesion Volume 5

Editor

K.L. Mittal

CRC Press
Taylor & Francis Group
Boca Raton London New York

CRC Press is an imprint of the
Taylor & Francis Group, an **informa** business

Contents

Contact Angle, Wettability and Adhesion, Vol. 5, pp. vii–viii
Ed. K.L. Mittal
© VSP 2008

Preface

This volume chronicles the proceedings of the Fifth International Symposium on Contact Angle, Wettability and Adhesion held under the aegis of MST Conferences in Toronto, Canada, June 21–23, 2006. The premier symposium with the same title was held in 1992 in honor of Prof. Robert J. Good as a part of the American Chemical Society meeting in San Francisco, California and the second, third and fourth events in this series were held in 2000 in Newark, New Jersey, 2002 in Providence, Rhode Island, and 2004 in Philadelphia, Pennsylvania, respectively. The proceedings of these earlier symposia have been properly documented as four hard-bound books [1–4]. Because of the high tempo of research and tremendous technological importance of this topic, the next (Sixth) symposium in this vein is planned to be held in Orono, Maine, July 14–16, 2008.

As pointed out in the Prefaces to the previous volumes, wettability is of pivotal importance in many and varied arenas, ranging from mundane (e.g., washing clothes) to high-tech (e.g., micro- and nanofluidics, lithography) to biomedical (e.g., contact lenses). It should be underscored that in the last years there has been burgeoning interest in replicating the so-called "Lotus Leaf Effect" to create superhydrophobic surfaces. A superhydrophobic surface is defined as the one with water contact angle greater than 150°, and it is based on the right combination of surface energy and surface topography. Even a cursory look at the literature will evince that recently there has been an accelerated pace of research in many and varied (of course, simple, practical and inexpensive) means to impart superhydrophobic character to a variety of surfaces – metallic, ceramic and polymeric. This high tempo of interest in these materials emanates from their applications, ranging from self-cleaning windows to controlling flow behavior in nanofluidics to tribological phenomena. Apropos, certain creatures (animals) capitalize on the superhydrophobic nature of certain parts of their body to collect water early in the morning (from dew) for subsequent use. Also it should be emphasized that there has been considerable interest in modulating the wetting characteristics of surfaces by electrical phenomena, the field known as "Electrowetting".

Now coming to this volume, it contains a total of 19 papers covering many facets of contact angle, wettability and adhesion. It must be recorded that all manuscripts were rigorously peer-reviewed and revised (some twice or thrice) and properly edited before inclusion in this book. Concomitantly, this volume represents an archival publication of the highest standard. It should not be considered a proceedings volume in the usual and ordinary sense, as many so-called proceedings are neither peer-reviewed nor adequately edited. By the way, the technical program for the symposium comprised many more papers, but those are not included in this volume for a variety of reasons (including some which could not pass muster).

This book (designated as Volume 5) is divided into three parts: Part 1: Contact Angle Measurements and Solid Surface Free Energy; Part 2: Relevance of Wetting in Cleaning and Adhesion; and Part 3: Superhydrophobic Surfaces. The topics covered include: Fundamental aspects of contact angle and its measurement; solidification contact angles of micro-droplets; microscopic wettability of wood cell walls; dynamic vapor–liquid interfacial tension; surface free energy of polymeric materials; surface cleanliness evaluation from wettability measurements; wettability parameters affecting surface cleanability of stainless steel and textiles; wetting and adhesion in fibrous materials; wettability and adhesion of coatings; adhesion of hydrophobizing agents; modulation of surface properties of polymers; graft efficiency and adhesion; relevance of interfacial free energy in cell adhesion; various approaches to create superhydrophobic surfaces; and adsorption of surfactants on hydrophobic and superhydrophobic surfaces.

This volume and its predecessors [1–4] containing bountiful information (a total of ~ 2900 pages) on various ramifications of contact angle, wettability and adhesion provide a comprehensive reference source and these volumes reflect the cumulative wisdom of a legion of active researchers engaged in understanding and harnessing the wettability phenomena. I sincerely hope these volumes would be of great interest and value to both neophytes (as a gateway to the field) as well as to veterans (as a commentary on contemporary research).

Acknowledgements

Now it is my pleasure to thank all those who helped in this endeavor. First, as usual, my heart-felt thanks go to Dr. Robert H. Lacombe, a dear friend and colleague, for taking care of the myriad details entailed in the organization of the symposium. Second, I would like to acknowledge the reviewers for their time and effort in providing many valuable comments which, most certainly, improved the quality of manuscripts contained in this volume. Special thanks are extended to the authors for their interest, cooperation and for providing written accounts of their presentations which formed the basis for this book. Finally my appreciation goes to the staff of Brill Academic Publishers for giving a body form to this book.

K. L. Mittal
P.O. Box 1280
Hopewell Jct., NY 12533

1. K. L. Mittal (Ed.), *Contact Angle, Wettability and Adhesion*, VSP, Utrecht (1993).
2. K. L. Mittal (Ed.), *Contact Angle, Wettability and Adhesion*, Vol. 2, VSP, Utrecht (2002).
3. K. L. Mittal (Ed.), *Contact Angle, Wettability and Adhesion*, Vol. 3, VSP, Utrecht (2003).
4. K. L. Mittal (Ed.), *Contact Angle, Wettability and Adhesion*, Vol. 4, VSP/Brill, Leiden (2006).

Part 1

Contact Angle Measurements
and Solid Surface Free Energy

Part I

Contact Angle, Wettability and Adhesion, Vol. 5, pp. 3–23
Ed. K.L. Mittal
© VSP 2008

Effects of surface structure on the behavior of a heated contact line

M. OJHA,[1] G. DALAKOS,[2] S. PANCHAMGAM,[1] P. C. WAYNER Jr.[1]
and J. L. PLAWSKY[1,*]

[1] *Dept of Chemical and Biological Engineering, Rensselaer Polytechnic Institute, Troy, NY 12180*
[2] *General Electric, Global Research Center, Niskayuna, NY 12309*

Abstract—The atomic force microscopy provides a detailed characterization of surface roughness and has enabled researchers to relate surface roughness to the macroscopic contact angle; however, only a little work has been done at the microscale, to tie surface parameters to the properties of the meniscus at the contact line. Here we report on studies looking at octane on a structured surface formed by a plasma etching process and focus on the effects of the surface on a heated contact line. We note that the adsorbed film thickness ahead of the contact line is a strong function of surface roughness, and so are the tangent angle and peak curvature at the contact line. The curvature exhibits a large jump in the region of the contact line which is a function of the surface roughness. The variation in adsorbed film thickness leads to an apparent Hamaker constant that is now a function of the surface structure as well as the dielectric properties of the medium. Peak curvature, tangent angle, and meniscus recession all are strong functions of heat input and surface roughness. As the surface roughens, the adsorbed film ahead of the meniscus becomes unstable, leading to a dropwise evaporation/condensation process.

Keywords: Surface roughness; AFM; RMS roughness; correlation length; sidewall meniscus; microscopic wetting; thickness profile; contact angle; tangent angle; curvature; heat transfer.

1. INTRODUCTION

Surface structure has a dominating influence at the three-phase contact line region where the interfacial parameters depend on the surface energies of the solid and the liquid phases. Controlling the interfacial force gradients that dominate contact line dynamics is critical to the success of technologies that involve boiling, isothermal and non-isothermal spreading and wetting, heat pipes, fuel cells, evaporation induced self-assembly, fluid distribution system in a lab-on-a-chip, and ink-jet printed rapid prototyping, etc. The future development and optimization of such technologies requires an understanding of the processes that occur at the solid-liquid interface. Such systems involve the presence of real surfaces that are not smooth.

---

*To whom correspondence should be addressed. Tel.: (518) 276-6049; Fax: (518) 276-4030;
e-mail: plawsky@rpi.edu

In these situations, contact line dynamics will be influenced by the structure of the solid surface.

The effects of surface structure on the contact angle and wettability of surfaces has a long history dating back more than 50 years to the pioneering work of Wenzel [1], Cassie and Baxter [2], Goode [3], and others [4–6]. Recently, the effects of surface roughness have become important as work on superhydrophobic surfaces by Miwa *et al.* [7] and patterned, switchable surfaces by Lahann *et al.* [8] has generated a great deal of interest. At the macroscopic scale, Wenzel [1] (Equation 2) and Cassie and Baxter [2] (Equation 3) developed theories that relate the apparent contact angle on a rough surface, θ_a, to the classical Young's contact angle, θ, (Equation 1) via the surface geometry represented by a roughness factor, r that is a ratio of surface areas. The Cassie-Baxter theory is an extension of Wenzel's theory and defines the apparent contact angle on a porous surface by introducing a porosity contribution, f_2.

$$\gamma_s = \gamma_{sl} + \gamma_l \cos\theta \tag{1}$$

$$\cos\theta_a = r\cos\theta; \quad r = roughness\ factor = \frac{actual\ surface\ area}{geometric\ surface\ area} \tag{2}$$

$$\cos\theta_a = r\cos\theta - f_2 \tag{3}$$

Wenzel's equation was experimentally verified on rough surfaces by Miller *et al.* [4] and Shibuichi *et al.* [5], however, the experimental data of Kawai and Nagata [6], Jopp *et al.* [9] and Fan *et al.* [10] on structured surfaces suggest that the prediction of the Wenzel equation lies below the observed values of the contact angle. Fan *et al.* [10] also found that when a composite air-solid layer formed below the drop, the increase in contact angle scaled well with the Cassie-Baxter equation.

These results have been explained by various researchers. According to Bico *et al.* [11] and Quere [12] the experimental apparent contact angle on structured surfaces is larger than that obtained from Wenzel's theory due to the presence of a thin film at the top of the porous solid that produces a "smoothing" effect on roughness. He *et al.* [13] proposed the presence of two energy states that follow the Wenzel and Cassie-Baxter equations, respectively. The geometric parameters of the underlying surface determine when the crossover occurs between the two energy states. Cheng *et al.* [14] have reported an observation of water condensation and evaporation on lotus leaf surfaces, where the roughness is comparable to the drop size. Cheng proposed that Young's model is valid locally on the surface and that the contact angle of water with respect to the leaf surface feature is small. However, the macroscopic contact angle with respect to the plane of the leaf appears large. According to the theoretical analysis of Palasantzas and De Hosson [15], the apparent contact angle depends critically on the roughness exponent, α, and the long wavelength ratio, w/ξ, of the surface. α, ξ and w are the statistical parameters that define the morphology of a rough surface. α is the micro-roughness parameter that defines the jaggedness of a surface, ξ is the equivalent of the wavelength of a rough surface; the surface morphology repeats itself beyond distance ξ. w represents the root mean square (RMS) roughness of the surface.

Most of the work published on surface roughness applies to partially wetting liquids. For a completely wetting film, the effect of surface roughness is seen in the adsorbed thin film. Garoff *et al.* [16] used x-ray reflectivity to show that the roughness of the solid affects the roughness of the liquid-vapor interface. A positive disjoining pressure favors thicker films that follow the surface geometry. However, the surface tension of the liquid film opposes the roughness in the film and favors a smooth film. Robbins *et al.* [17] proposed a healing length, Y, (Eq. 4) beyond which the dispersion and structural contributions to the disjoining forces fall off and the liquid-vapor interface becomes smooth.

$$Y = \frac{\delta^2}{a}, \qquad a = \left(\frac{A}{2\pi\gamma}\right)^{\frac{1}{2}} \qquad (4)$$

where, δ is the mean height (average height of the liquid); Y is the healing length; A is the Hamaker constant and γ is the liquid surface tension.

Neither Wenzel and Cassie-Baxter theories nor the work of Garoff *et al.* [16] or Robbins *et al.* [17] provide any conclusive evidence on the effect of the structured surfaces on any of the "microscopic" interfacial parameters, like the thickness profile, tangent angle profile and the curvature profile. Moreover, available theories are applicable only to the isothermal conditions and, therefore, the effect of surface structure on interfacial profiles for heated surfaces needs to be studied to obtain further insight into the transport processes that occur at the contact line under those conditions.

The relationship between interfacial parameters and the transport properties at the contact line on a smooth surface has been widely studied in the past. On smooth surfaces, DasGupta *et al.* [18] have shown that the experimentally measured shape of the liquid-vapor interface at the micro-scale, represented by the film thickness, the tangent angle and the curvature, gives the isothermal interfacial free energy per unit volume, or the pressure jump at the interface. Derjaguin and Zorin [19], Potash and Wayner [20], Moosman and Homsy [21], Wayner [22], and Gokhale *et al.* [23] have shown that the change in the chemical potential per unit volume of a single component liquid in a gravitational field, $\Delta\mu_g$, is a function of pressure and temperature jumps across the liquid-vapor interface, which is predicted by Equation (5).

$$\Delta\mu_g = -(K\gamma + \Pi) + \frac{\rho_l \Delta H_m}{T}(T_{lv} - T_v) \qquad (5)$$

where, ρ_l is the liquid density (kg/m^3), K is the interfacial curvature, γ is the interfacial tension, Π is the disjoining pressure, and ΔH_m is the heat of vaporization. The interfacial temperature jump is $\Delta T = T_{lv} - T_v$. Using Equation (5) for the completely wetting isothermal case, equilibrium ($\Delta\mu_g = 0$; $\Delta T = 0$) results can be obtained experimentally and an *in-situ* effective Hamaker constant, A_{eff}, can be determined. The effective Hamaker constant is not only a function of solid and liquid properties but also includes the effect of surface roughness. For isothermal equilibrium, the extended Young-Laplace equation is obtained (Equation (6)). The interfacial vapor pressure $(P_v - P_l)$ is affected by capillarity $(K\gamma)$, disjoining pressure (Π) and temperature.

$$K\gamma + \Pi = P_v - P_l; \qquad \Pi = \frac{A_{eff}}{6\pi\delta^3} \tag{6}$$

Wayner and co-workers (Wayner *et al.* [22, 24], Renk *et al.* [25, 26], Liu *et al.* [27], DasGupta *et al.* [28]) have used the augmented Young-Laplace model, which includes the concept of disjoining pressure for the interfacial force field characterization of a thin film, together with an interferometry technique, to study the interline region of an evaporating, wetting meniscus. They concluded that the shape of the meniscus, specifically the curvature, influenced the heat flux near the contact line region and that the pressure gradient for flow was associated with gradients in the meniscus curvature and thickness.

The surface morphology of an irregular surface would influence the microscopic shape of the liquid-vapor interface. In this study we present experimental evidence for the effect of the solid surface microstructure on the interfacial shape of the liquid under isothermal and heated conditions. Here, we have used an optical interferometry technique to obtain the shape profile in the vicinity of the contact line. A higher resolution version of the optical interferometry technique enables the measurement of meniscus profile at the microscopic scale where the intermolecular force field is dominant. The solid structure was accurately analyzed by means of Atomic Force Microscopy (AFM). A height-height correlation function, $H(r)$, was used to determine the statistical features of the rough surface. Thus, a relationship between the surface morphology parameters and the wetting properties of the octane sidewall meniscus was obtained. We show that the adsorbed film thickness, tangent angle and the curvature profile which represents the shape of the liquid region during the microscale phase change process, all were a strong function of the solid surface structure during isothermal as well as non-isothermal conditions. Modification in the surface potential as a result of different interfacial profiles on the irregular surface is also demonstrated. The resulting shape of the sidewall meniscus, which is a function of the surface morphology, led to different transport properties of the sidewall meniscus which were seen during meniscus recession under non-isothermal conditions.

2. EXPERIMENTAL SET-UP

An experimental set-up was designed for a completely or partially wetting apolar liquid system where the intermolecular force fields can be easily modeled and the effects of the surface structure can be studied by means of a simple set-up. The set-up under study has a constrained vapor bubble (CVB) design. A channel of $65 \times 14 \times 3$ mm dimensions was etched inside an aluminum plate. The periphery of the channel was surrounded by a groove where an O-ring was placed to seal the system. The channel was filled with the liquid and covered at the top by a transparent solid surface. The O-ring filled the space between the channel and the microscope slide and clamps were used to completely seal the system. Once the channel was filled with the liquid, the liquid evaporated from the bottom of the channel and condensed on the top surface. After condensation, the liquid formed a thin adsorbed film on the rough surface and a sidewall meniscus was formed at the corners. As a result, a vapor bubble was enclosed in the cell and was surrounded by menisci that were formed at the corners and an adsorbed thin film that formed on the substrate surface.

A simple schematic of the CVB and the liquid-solid system is shown in Fig. 1(c). All the regions inside the cell were connected by ultra-thin liquid films and through the vapor phase. The rough surface under study was fabricated on the inside of the solid substrate so that it was in contact with the liquid in the channel. The cell was kept at a slight tilt relative to the horizontal so that a liquid reservoir was present at the lower end of the cell (Fig. 1(b)). A heater was attached to the outside surface of the quartz substrate at the higher end to enable experiments under non-isothermal conditions. The sidewall meniscus region and the adsorbed liquid film

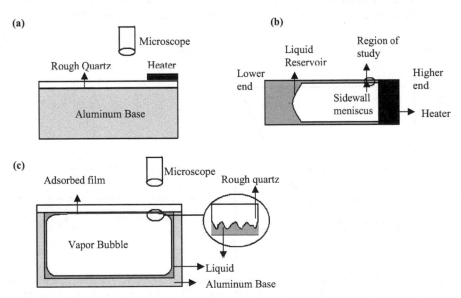

Figure 1. Schematic drawing of (a) the cross section, (b) Side view and (c) Top view of the experimental setup. Rough quartz slide is kept upside down on a channel filled with liquid. Liquid condenses on the quartz to form an adsorbed film.

M. Ojha et al.

that formed between the liquid and the rough solid channel cover were easily viewed from the top through a microscope. The observed image was then recorded for further analysis.

Of particular interest was the contact line region formed at the solid-liquid-vapor interface. For a sidewall meniscus there is an associated thin film in the microscopic region ahead of the meniscus. The macroscopic region is defined here as the region of film thickness larger than $\delta = 0.1$ μm. The interfacial region below this film thickness is defined here as the microscopic region and so for convenience, the contact line is defined as the interface between the two regions. A schematic drawing of the interfacial profile in the microscopic region is sketched in Fig. 2. The contact line region extends from a thickness of a few molecular diameters to the macroscopic region.

Upon shining light through a constantly changing film thickness of the meniscus, an interference fringe pattern is produced. A typical interference pattern of octane sidewall meniscus on a rough substrate is shown in Fig. 2(b). The constant thickness of the adsorbed film produces a constant reflectivity, as seen on the left side of the figure. Single beam interferometry was used to analyze liquid profiles of the drops or the meniscus. Monochromatic green light from a Hg-arc lamp at a wavelength of 546 nm was used to illuminate the liquid-solid interface. Reflection of light at the liquid-vapor and liquid-solid interfaces produced interference fringes. A CCD camera was used to capture the images of interference fringes. Commercially available software, Image-pro Plus® (MediaCybernetics, Bethesda, MD), was used

(a)   **(b)**

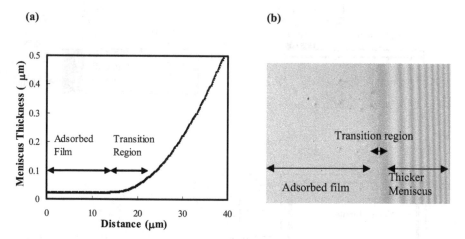

Figure 2. (a) Typical thickness profile obtained for the sidewall meniscus. Adsorbed film thickness is constant in the adsorbed film region shown in the figure. The region between the constant thickness and 0.1 μm thickness is marked as transition region. (b) Typical interference fringes obtained from the constantly changing thickness of the sidewall meniscus. Approximate locations of the adsorbed film region, transition region and thicker meniscus are also shown.

Table 1.
Physical properties of n-octane

Dielectric Constant, ε	Refractive index, n	Boiling point, °C	Surface tension, σ, mN/m	Dynamic viscosity, μ $\times 10^3$ Pa.s	Density, ρ, kg/m^3	Heat of vap., ΔH_m, kJ/kg
1.95	1.398	126	21.1	0.508	703	363

to analyze the recorded images from which a gray value plot of the interference fringes was obtained. The interference pattern was analyzed to obtain a thickness profile by using the technique described by Gokhale *et al.* [29]. The obtained thickness profile was then fitted with a Matlab-based spline function to evaluate its first and second derivatives and obtain the tangent angle and curvature profiles, respectively, by using a fitting procedure employed by Panchamgam *et al.* [30]. The physical properties of the working fluid octane used for various calculations in this paper are listed in Table 1.

Quartz microscope slides were used as the solid substrates. Polished quartz slides manufactured by Technical Glass Products, Inc (Painesville Twp., OH) have an RMS roughness, w, of ~0.5 nm and these were selected as a reference for smooth quartz. Quartz substrates manufactured by Electron Microscopy Sciences (Hatfield, PA) are not heavily polished. The RMS roughness of these substrates was ~1.0 nm. These microscope slides were etched in fluorocarbon plasma to generate samples with even higher roughness. Plasma etching of the quartz slides was performed in a commercially available 13.56 MHz Plasma-Therm 73 (PlasmaTherm Inc, Voorhees, NJ) capacitively-coupled etching reactor. An O_2(18%)/CF_4 feedstock gas mixture was introduced through a top "showerhead" electrode. RF power was supplied to the bottom electrode while the top electrode was grounded. The bottom electrode was water cooled to 25°C. Clean substrates were placed at the center of the bottom electrode. After a vacuum pump-down, argon was flowed over the samples to purge the reactor before etching. The following processing parameters were used to etch the samples – 200 W/200 mTorr/40 SCCM O_2+CF_4. The powered electrode and the presence of O_2 in the feedstock gas prevents formation of any polymeric by-products during plasma etching. Following etching, the quartz samples were cleaned using a piranha solution (66% H_2SO_4 + 34% H_2O_2) for 20 minutes to remove any organic impurities. Any residue left as a result of the piranha cleaning was removed by repeatedly rinsing with distilled water. Finally, the substrates were dried with dry N_2 gas to remove any traces of water.

Surface roughness characterization of the quartz microscope slides was done using AFM scans. $5 \times 5\,\mu m^2$ area scans of the substrates were taken under intermittent contact mode using ParkScientific-CP® scanning probe unit with Ultralever® cantilevers. A height-height correlation function, $H(r)$, was used to obtain the parameters that statistically define the surface morphology, namely, the correlation length, ξ, the RMS roughness, w, and the roughness exponent, α. $H(r)$

M. Ojha et al.

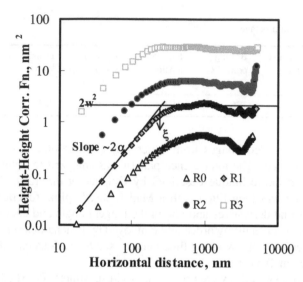

Figure 3. Height-Height correlation function obtained for four quartz surfaces. The slope of each curve yields the microroughness parameter, α, the saturation of each curve yields RMS roughness, w, and the point of intersection between the two curves yields the correlation length, ξ. Surface morphology parameters α, w and ξ were obtained from this height-height correlation function for all quartz surfaces.

Table 2.
Surface morphology parameters obtained for quartz surfaces

	α	w, nm	ξ, nm	w/ξ	R_{p-v}, nm	R_{p-v}/w
R0	0.64	0.5	404	1.3E-3	3.4	6.5
R1	0.87	1.0	253	4.3E-3	8.1	7.5
R2	0.80	1.7	195	9.0E-3	11	6.2
R3	0.71	3.7	161	2.3E-2	26	6.9

is related to these functions as, $H(r) = \rho^2 r^{2\alpha}$ for $r \ll \xi$ and $H(r) = 2w^2$ for $r \gg \xi$, where ρ is the average local slope. Plots of $H(r)$ obtained for our surfaces are shown in Fig. 3. From the above equations it can be seen that the slope of the curve is $\sim 2\alpha$, the saturation value of $H(r)$ represents w and the crossover point represents ξ.

The height-height correlation functions obtained for all the surfaces are shown in Fig. 3. The polished quartz manufactured by Technical Glass Products Inc, is called R0, whereas the unpolished quartz obtained from Electron Microscopy Sciences is called R1. Two samples of rough quartz were produced from plasma etching of R1, and these are called R2 and R3, respectively. The roughness parameters are shown in Table 2 and the surface scans obtained from the AFM are shown in Fig. 4. R0 was

(a)   (b)

(c)

(d)

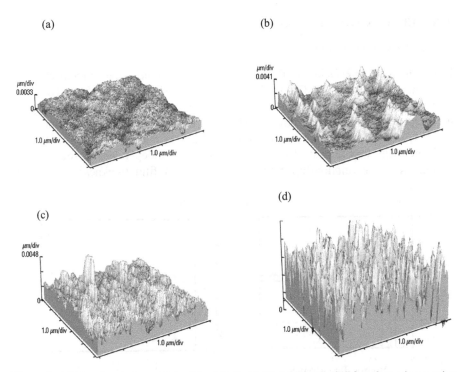

Figure 4. AFM surface images obtained for (a) R0, (b) R1, (c) R2 and (d) R3 surfaces via a tapping mode scan.

found to have a low α as well as a low w. R1 had a higher roughness which arises from the isolated hillock-like structures that can be seen in its AFM image. Thus, this surface is mostly smooth but has random hillock-like features. We have used the ratio of peak to valley distance, R_{p-v}/w, to define such a feature. It can be seen that R_{p-v}/w is highest for the unpolished quartz (R1). Also, the micro-roughness represented by α was highest for the unpolished quartz. After R1 was etched, the hillocks were also etched and the peaks and valleys were closer and more uniformly distributed. This effect can be seen as both, ξ and R_{p-v}/w decreased on R2 and R3. However, w was found to be a monotonically increasing function of the etch time and thus could be controlled precisely by varying the etch time. ξ decreased when R1 quartz was etched but it remained independent of the etch time. The etched microscope slides that were used to analyze the interfacial profiles by our optical interferometry method were all transparent at 546 nm and the order of roughness was too small ($w < 5\,\mathrm{nm}$) compared to the wavelength of light for scattering to be important.

3. RESULTS AND DISCUSSION

3.1. Isothermal conditions

Isothermal state interfacial profiles were studied for octane sidewall menisci on the rough quartz surfaces. The film thickness profiles are shown in Fig. 5(a). The effect of roughness was most prominent in the adsorbed film and the transition region, the shape of those regions is critical to the heat transfer properties of the meniscus [22, 24]. Under isothermal "equilibrium" conditions, the thickness profile in the bulk meniscus was almost identical on all surfaces, independent of the substrate roughness. The relationship between the adsorbed film thickness, w and the

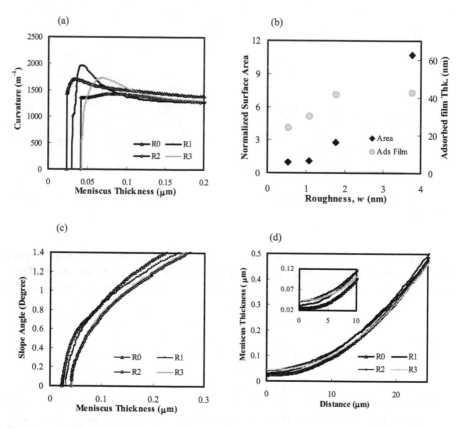

Figure 5. (a) Meniscus thickness profile obtained under isothermal conditions. Thickness profile beyond the adsorbed film region where transition region starts is shown. (b) Adsorbed film thickness as a function of surface roughness. Adsorbed film thickness initially increases monotonically as the roughness is increased but eventually saturates at higher surface roughness. (c) Curvature profiles obtained during isothermal equilibrium condition. (d) Tangent angle profiles obtained on various surfaces during isothermal equilibrium condition.

normalized surface area (actual surface area/projection of the surface area) is shown in Fig. 5(b). The adsorbed film thickness increased monotonically with increasing w at lower roughness values but eventually saturated as w increased beyond $\sim$2 nm. The increase in adsorbed film thickness with w has been theoretically predicted by Robbins *et al.* [17], who state that the roughness of the solid decreases the disjoining pressure at a fixed chemical potential which increases the average adsorbed film thickness.   Also, the actual surface area of a rough surface is greater than its projected area. Palasantzas [31] has derived Equation (7) that provides the upper limit of the increased surface area due to roughness. A normalized surface area, relative to the polished glass slide, was obtained using the expression provided by Palasantzas. Before the adsorbed film saturates, an increase in the surface area is correlated to an increase in the adsorbed film thickness.

$$K_1^2 = \frac{w^2}{2a^2\xi^2}\left\{\frac{1}{1-\alpha}[(1+aQ_c^2\xi^2)^{1-\alpha}-1]-2a\right\} \quad \text{for } 0 \leqslant \alpha < 1 \qquad (7)$$

$$a = \frac{1}{2\alpha}\left[1-(1+aQ_c^2\xi^2)^{-\alpha}\right]$$

$$Q_c = \frac{\pi}{a_0}; \quad \text{where, } a_0 \text{ is the inter-atomic spacing}$$

where, K_1 is the surface area normalized with the projected area.

   Wenzel [1] and Palasantzas and De Hosson [15] predict that the contact angle is a function of surface roughness parameters, namely, w, α and w/ξ. The tangent angle profiles obtained on various surfaces are shown in Fig. 5(d). We found a decrease in isothermal state tangent angle profiles with an increase in roughness, w, as well as with w/ξ as theoretically predicted by Palasantzas and De Hosson. However, they also predict that the contact angle is a function of α. Our results for the tangent angle profile bear no relationship with α. Wenzel, and Palasantzas and De Hosson theories predict that the contact angle should decrease with roughness of the surface; this trend is also seen in our results. However, the equations proposed by Wenzel or by Palasantzas and De Hosson to obtain the apparent contact angle from the surface morphology parameters are applicable only to partially wetting systems where the contact angle is high. For our completely wetting sidewall meniscus, both theories break down as they predict non-physical contact angles.

   Isothermal state curvature profiles are shown in Fig. 5(c). The profiles are identical in the bulk meniscus region indicating that the effect of surface roughness is negligible at higher film thicknesses. In the transition region, small jumps are observed in the curvature profiles. The curvature profile obtained for sample R1 shows the highest jump. The AFM scan of this surface shows isolated hillocks on a smoother surface and the presence of such hillocks is reflected in the curvature profile in the transition region. The ratio of R_{p-v}/w is a high number when R_{p-v} is large and w is small, therefore for a surface with isolated hillocks (large R_{p-v}) on a relatively smooth surface (small w) we should see a direct relationship between the

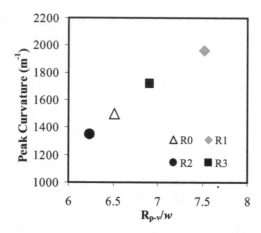

Figure 6. Peak curvature vs R_{p-v}/w. Under isothermal conditions peak curvature increases linearly as R_{p-v}/w is increased.

curvature jump and this surface feature measure. Figure 6 shows that the curvature peak increases monotonically with R_{p-v}/w.

3.2. Non-isothermal conditions

A constant heat flux was applied to the sidewall meniscus to study its behavior under non-isothermal conditions. At $t = 0$ seconds the heater power was abruptly increased to the desired level and kept constant thereafter. As the heat flux was applied the meniscus began to recede. Steady-state was achieved once the recession ended. The distance traveled during recession and the non-isothermal steady-state of the sidewall meniscus were studied at four different heater powers of 0.18 W, 0.8 W, 1.8 W and 3.36 W. At lower heater powers the meniscus was stable on all the surfaces. However, meniscus instability appeared on the R3 quartz at higher heater powers (0.8 W, 1.8 W, 3.36 W) leading to the formation of drops ahead of the actual meniscus.

Figure 7(a) shows the meniscus thickness as a function of distance at a heater power of 0.18 W. All profiles show the presence of a thin adsorbed film of a constant thickness in front of the meniscus. In Fig. 7(a), the start of the x-axis (distance) is set at the first data point beyond which the film thickness increases or the location where the adsorbed film thickness ceases to exist and the transition region begins. Application of a heat flux caused a reduction in the adsorbed film thickness. On all surfaces, the adsorbed film thickness was reduced from its isothermal value. At steady-state, the adsorbed film thickness was higher when the substrate roughness was higher, similar to what was observed under isothermal conditions. It can be seen from Fig. 7(a) that the spread of the meniscus under non-isothermal conditions was entirely different than that seen under isothermal conditions. The meniscus spreads

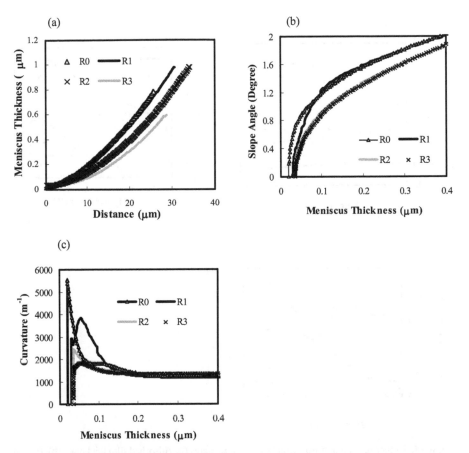

Figure 7. (a) Meniscus thickness profiles obtained on all surfaces under non-isothermal equilibrium condition at a heater power of 0.18 W. Meniscus spreads more on rougher surfaces. (b) Tangent angle profiles obtained on all surfaces under non-isothermal equilibrium condition at a heater power of 0.18 W. Tangent angle reduces on surfaces with higher roughness. (c) Curvature profiles obtained during non-isothermal equilibrium condition when 0.18 W heater power was applied. Peak in the curvature profile decreases as surface roughness is increased.

more on the substrates with higher roughness, as expected. Roughness effects were also reflected in the tangent angle (Fig. 7(b)) and the curvature profiles (Fig. 7(c)). The tangent angle profile was found to be decreased on the surfaces that exhibited higher roughness. The curvature profiles exhibited a distinct peak in the transition region that increased relative to the isothermal state, but decreased with increasing surface roughness.

Qualitatively similar behavior was observed at higher heater powers. Comparisons between the adsorbed film thickness, tangent angle at a film thickness of

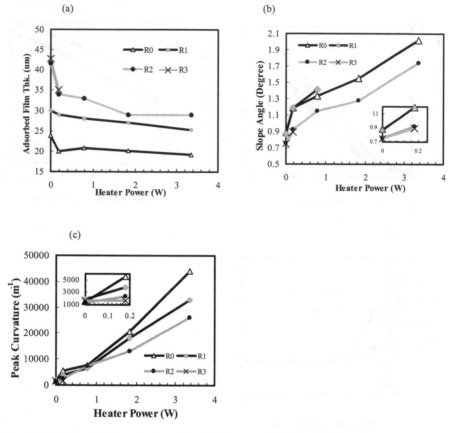

Figure 8. Isothermal and non-isothermal equilibrium states (a) Adsorbed film thickness, (b) Tangent angle at a meniscus thickness of 100 nm, and (c) The peak curvature obtained on different surfaces at various heater powers.

100 nm and the peak curvature observed on all surfaces at every heater power under non-isothermal equilibrium are shown in Figs. 8(a), (b) and (c) respectively. On each individual surface the adsorbed film thickness at lower heater powers decreased from its isothermal state value but was found to eventually saturate when higher heater power was applied. On each individual surface, the meniscus tangent angle and the peak curvature consistently increased as the heater power was increased. The effect of evaporation on the contact angle of a spreading drop of a wetting liquid on a heated surface has been solved theoretically by Ajaev [32]. According to this model, droplet spreading is governed by a competition between evaporation and capillary spreading where evaporation prevents spreading, and as a result the observed contact angle is higher.

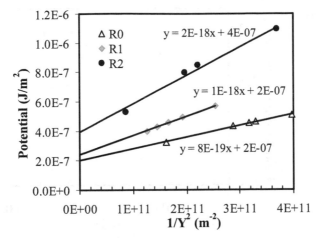

Figure 9. Relationship between the healing length, Y, and the effective potential observed on a rough surface. Intercept on the y-axis represents the potential on a flat substrate.

The adsorbed film thickness determines the surface potential for a liquid-solid system. Since the adsorbed film thickness is a function of roughness, this poses an interesting problem as to how to define the surface potential on a rough surface. Robbins *et al.* [17] and Palasantzas [33] have addressed this and presented Equation (8) which relates healing length, Y, and surface morphology parameters, α, ξ and w to the surface potential. Y is a function of mean film thickness, Hamaker constant and surface tension as shown in Equation (4). The surface potential for our substrates, $U_e(\varepsilon)$, was calculated from Equation (9). The effective Hamaker constant, A_{eff}, was obtained from Equation (6) for a steady-state isothermal case where the disjoining and the capillary terms are equal. Thus, the effective Hamaker constant was obtained experimentally. Figure 9 shows a plot between $U_e(\varepsilon)$ and $1/Y^2$ and straight lines result for all surfaces. The slope of each line represents Ω, which is a function of the surface morphology, and the intercept of each line represents the effective potential on a flat surface. As the surface becomes rougher the effective potential increases. Results from Fig. 9 follow this trend and all the surfaces show a similar intercept on the y-axis which indicates that the effective potential on a flat surface is a constant number. Also, the intercept for each surface is close to the potential calculated for R0, which is the smooth surface.

$$U_e(\varepsilon) \approx U(\varepsilon) + \Omega Y^{-2} \quad (0 \leqslant \alpha < 1, \ \xi \ll Y) \tag{8}$$

where, $\Omega \approx \{[K\sigma^2 \ln(Q_c^2\xi^2)]/a(4\pi^2)\}$

$$U_e(\varepsilon) = \frac{A}{12\pi\delta_0^2} \tag{9}$$

M. Ojha et al.

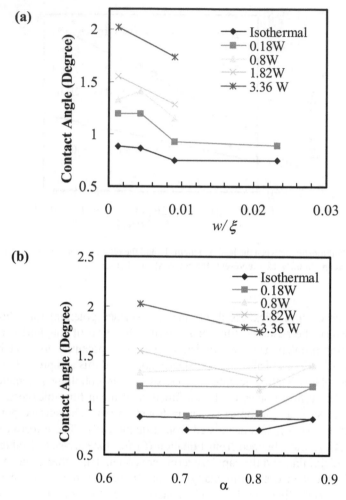

Figure 10. Tangent angle at 100 nm meniscus thicknesses as a function of (a) w/ξ and (b) α. Tangent angle decreases with increasing w/ξ. However, no direct relation was observed between tangent angle and α.

The adsorbed film thickness was higher and the tangent angle (Fig. 8) was lower for rougher surfaces at all heat fluxes. We found that the tangent angle was a strong function of the long wavelength ratio, w/ξ, on all the surfaces even under non-isothermal conditions as shown in Fig. 10(a). Such a result has been theoretically predicted by Palasantzas and De Hosson [15] for the contact angle of drops on rough surfaces under isothermal conditions. However, the theory also predicts that the contact angle is a function of microscopic roughness parameter, α. As seen from

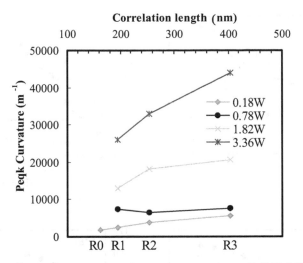

Figure 11. Peak curvature on all rough surfaces at different heat inputs (0.18 W, 0.78 W, 1.82 W and 3.36 W) as a function of correlation length of the rough surface.

Fig. 10(b) our results do not show any direct relationship between α and the tangent angle.

The tangent angle and the peak curvature depict the spread of the meniscus. The peak curvature was found to be higher on smoother surfaces. Moreover, the difference between the peak curvatures obtained on the smooth and the rough surface increased as the heater power was increased. Robbins *et al.* [17] and Palasantzas [33] have shown that surface roughness generates an additional local curvature in the thin adsorbed film. The effective potential now depends on the disjoining pressure as well as the local curvature. This local curvature is a function of ξ of the solid surface. The relationship between the observed peak curvature (transition region) and ξ is shown in Fig. 11. The peak curvature was found to be a function of ξ. Higher curvature was observed on surfaces with a higher ξ.

A step change in the heater power caused evaporation from the meniscus. Evaporation from the meniscus was seen in the form of meniscus recession on the solid surface. Recession curves, measured from the motion of the first dark fringe ($\delta = 100$ nm) for all surfaces at different heat fluxes are shown in Fig. 12. The plots of R0 and the R1 surfaces show similar behavior for all heater powers. The total distance that the meniscus receded at each heater power, i.e. at the steady-state, is shown in Fig. 13. At lower heater powers (0.18 W, 0.8 W) the recession on smooth surface, R0, was higher in comparison to the etched surface R2. At higher heater powers a crossover occurred and recession on R0 was less compared to that obtained on R2. Higher recession on the rough surface would imply that more evaporation had occurred. The non-isothermal steady state meniscus thickness profiles (Fig. 7(a)) show more spread of the meniscus on a rough surface. This also

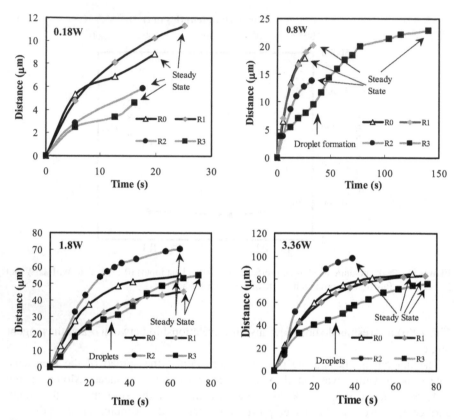

Figure 12. Distance travelled by the first dark film as a function of time at various heater powers.

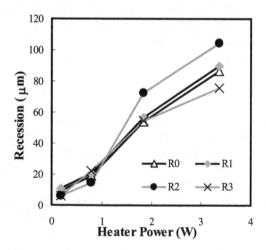

Figure 13. Meniscus recession at different heater powers.

(a) (b)

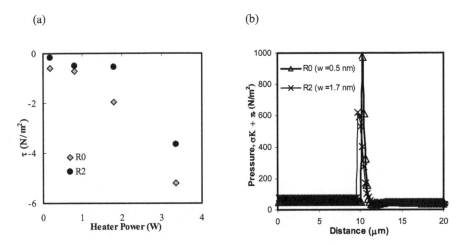

Figure 14. (a) Calculated shear stress, τ, as a function of heater power applied. Difference in stresses on the two surfaces increases as the heater power is increased. (b) Pressure profiles, $\sigma K + \pi$, obtained on R0 and R2 at a heater power of 3.36 W.

implies a larger transition region on the rough surface. With the exception of the adsorbed film region, the volume included in the meniscus on a rough surface at the steady state non-isothermal condition was less than that on the smooth surface. To dissipate the same power on a smooth surface more liquid flow was required towards the smaller transition region of the rough surface. The curvature and thickness profiles were used to obtain the pressure difference driving the flow toward the adsorbed film region. The pressure profiles shown in Fig. 14(b) show that the pressure driving force observed on R0 was much higher than that observed on R2. A higher pressure difference implies more liquid flow towards the transition region.

An average value for the shear stress acting on the liquid by the solid was calculated using the approximate model of Panchamgam *et al.* [34]. The control volume was chosen as the region where the liquid meniscus thickness increases from the adsorbed film thickness until it reaches 300 nm. The results obtained from the shear stress model for all the heat fluxes on R0 and R2 are shown in Fig. 14(a). At a lower heat flux the shear stress experienced by the liquid meniscus on both surfaces was almost equal. However, at a higher heat flux, R0 experienced a higher shear stress in comparison to R2. The sign of the stress was negative meaning that the flow was towards the adsorbed film region. More liquid flow towards the adsorbed film region of R0 results in a higher shear stress and there was more resistance to the liquid flow on a smooth surface.

At even higher roughness, on R3, the meniscus movement at the lowest heater power (0.18 W) was in accordance with the above discussion. At higher heater powers (0.8 W, 1.8 W, 3.36 W) the transition region broke down and led to instability. The instability occurs in the form of droplet formation in front of the

Figure 15. Instability observed on the rough surface. Image of droplet instability which forms in front of the receding meniscus on R3 at a heater power of 0.8 W.

receding meniscus as shown in Fig. 15. The event of droplet formation probably retarded the meniscus recession (seen as a change in the shape of the recession plot). Such a behavior was observed consistently for all higher heater powers. Droplet formation could be a result of breakup of the adsorbed film in front of the meniscus or due to distillation of impurities or probably some other reason. A detailed investigation on the occurrence of instability and its role in meniscus recession is beyond the scope of this article and this will be carried on as a follow-up to the work presented here.

4. CONCLUSIONS

The effect of surface roughness on the resulting shape of the sidewall meniscus at the liquid-solid interface was studied. The surface morphology parameters obtained from the height-height correlation function were correlated to the interfacial properties, namely the thickness profile, tangent angle and the curvature profile. Roughness enhances the wetting properties of the meniscus under isothermal as well as under non-isothermal conditions. The most significant effect of roughness was seen on the adsorbed film thickness and the curvature profile. Adsorbed film thickness was found to be higher on rough surfaces; the curvature profile exhibited a lower peak in the transition region as the roughness was increased. The effective potential was also found to increase as the surface roughness increased and to be inversely related to the square of the healing length as predicted by previous theories. Enhanced spreading of the meniscus on a rough surface led to lower curvature gradients and reduced pressure differential for the liquid flow towards the transition region. Higher liquid flow on a smooth surface leads to higher resistance. Enhanced spreading and low liquid flow resistance on rough surfaces might offer an advantage in the heat transfer efficiency of the meniscus. At even higher surface roughness, meniscus instability was observed in the form of droplet formation and undulations in the contact line. A detailed investigation of the instabilities will be carried out in the future to completely understand the effect of roughness on contact line behavior.

REFERENCES

1. R. N. Wenzel, *Ind. Eng. Chem.*, **28**, 988 (1936).
2. A. B. D. Cassie and S. Baxter, *Trans. Faraday Soc.*, **54**, 546 (1994).
3. R. J. Good, *J. Am. Chem. Soc.*, **74**, 504 (1952).
4. J. D. Miller, S. Veeramasuneni, J. Drelich, M. R. Yalamanchili and G. Yamauchi, *Polym. Eng. Sci.*, **36**, 1849 (1996).
5. S. Shibuichi, T. Onda, N. Satoh and K. Tsujii, *J. Phys. Chem.*, **100**, 19512 (1996).
6. A. Kawai and H. Nagata, *Jpn. J. Appl. Phys.*, **33**, 1283 (1994).
7. M. Miwa, A. Nakajima, A. Fujishima, K. Hashimoto and T. Watanabe, *Langmuir*, **16**, 5754 (2000).
8. J. Lahann, S. Mitragotri, T.-N. Tran, H. Kaido, J. Sundaram, I. S. Choi, S. Hoffer, G. A. Somorjai and R. Langer, *Science*, **299**, 371 (2003).
9. J. Jopp, H. Grull and R. Yerushalmi-Rozen, *Langmuir*, **20**, 10015 (2004).
10. J.-G. Fan, X.-J. Tang and Y.-P. Zhao, *Nanotechnology*, **15**, 501 (2004).
11. J. Bico, U. Thiele and D. Quere, *Colloids Surfaces A*, **206**, 41 (2002).
12. D. Quere, *Physica A*, **313**, 32 (2002).
13. B. He, N. A. Patankar and J. Lee, *Langmuir*, **19**, 4999 (2003).
14. Y.-T. Cheng, D. E. Rodak, A. Angelopoulos and T. Gacek, *Appl. Phys. Lett.*, **87**, 194112 (2005).
15. G. Palasantzas and J. T. M. De Hosson, *Acta Mater.*, **49**, 3533 (2001).
16. S. Garoff, E. B. Sirota, S. K. Sinha and H. B. Stanley, *J. Chem. Phys.*, **90**, 7505 (1989).
17. M. O. Robbins, D. Andelman and J.-Fo. Joanny, *Physical Review A*, **43**, 4344 (1991).
18. S. DasGupta, I. Y. Kim and P. C. Wayner, Jr., *AIChE J.*, **41**, 2140 (1995).
19. B. V. Derjaguin and Z. M. Zorin, *Proc. 2nd Intl. Congr. Surface Activity*, J. H. Schulman (Ed.) Vol. 2, p. 145, Butterworth, London (1957).
20. M. Potash, Jr. and P. C. Wayner, Jr., *Int. J. Heat Mass Transfer*, **15**, 1851 (1972).
21. S. Moosman and G. M. Homsy, *J. Colloid Interface Sci.*, **73**, 212 (1980).
22. P. C. Wayner, Jr., *Colloids Surfaces*, **52**, 71 (1991).
23. S. J. Gokhale, J. L. Plawsky and P. C. Wayner, Jr., *J. Colloid Interface Sci.*, **259**, 354 (2003).
24. P. C. Wayner, Jr., Y. K. Kao and L. V. LaCroix, *Int. J. Heat Mass Transfer*, **19**, 487 (1976).
25. F. Renk, P. C. Wayner, Jr. and G. M. Homsy, *J. Colloid Interface Sci.*, **67**, 408 (1978).
26. F. J. Renk and P. C. Wayner, Jr., *J. Heat Transfer*, **101**, 55 (1979).
27. A.-H. Liu, P. C. Wayner, Jr. and J. L. Plawsky, *Phys. Fluids*, **6**, 1963 (1994).
28. S. DasGupta, I. Y. Kim and P. C. Wayner, Jr., *J. Heat Transfer*, **116**, 1007 (1994).
29. S. J. Gokhale, J. L. Plawsky, P. C. Wayner, Jr. and S. DasGupta, *Phys. Fluids*, **16**, 1942 (2004).
30. S. S. Panchamgam, J. L. Plawsky and P. C. Wayner, Jr., *J. Heat Transfer*, **128**, 1266 (2006).
31. G. Palasantzas, *J. Appl. Phys.*, **81**, 246 (1997).
32. V. S. Ajaev, *J. Fluid Mech.*, **528**, 279 (2005).
33. G. Palasantzas, *Physical Review B*, **51**, 14612 (1995).
34. S. S. Panchamgam, S. J. Gokhale, J. L. Plawsky, S. DasGupta and P. C. Wayner, Jr., *J. Heat Transfer*, **127**, 231 (2005).

Contact Angle, Wettability and Adhesion, Vol. 5, pp. 25–46
Ed. K.L. Mittal
© VSP 2008

Apparent solidification contact angles of micro-droplets deposited on solid surfaces

RI LI,[1] NASSER ASHGRIZ,[1,*] SANJEEV CHANDRA,[1]
JOHN R. ANDREWS[2] and STEPHAN DRAPPEL[3]

[1]*Department of Mechanical and Industrial Engineering, University of Toronto, 5 King's College Road, Toronto, Ontario M5S 3G8, Canada*
[2]*Xerox Corporation, Wilson Center for Research & Technology, 800 Phillips Rd. M/S 114-44D, Webster, NY 14580*
[3]*Xerox Research Centre of Canada, 2660 Speakman Drive, Mississauga, Ontario L5K 2L1, Canada*

Abstract—An experimental investigation of apparent solidification contact angle formed in deposition of ink droplets on solid surfaces as occurs in solid inkjet printers is presented. The apparent solidification contact angles of 39 μm diameter droplets of hot-melt ink impacting on sub-cooled solid surfaces under different printing conditions are obtained. The printing conditions were varied by varying four parameters: the type of substrate, the distance between the substrate surface and printhead, the substrate temperature, and the printhead jetting temperature. It is found that the apparent contact angle is not only dependent on the substrate material and substrate temperature but also has strong dependence on the impact velocity and temperature of the droplet. Explanation is provided by considering the coupling effect of viscous damping and impact process.

Keywords: Contact angle; droplet impact; solidification; heat transfer; hot-melt ink.

Nomenclature

C_p	specific heat
D	diameter of droplet
g	gravitational acceleration
k	thermal conductivity
k_d	thermal conductivity of droplet
k_s	thermal conductivity of substrate
l	thickness
L	flight distance
L_f	latent heat
h	height of deposited droplet
m	mass

*To whom correspondence should be addressed. Tel.: 416-946-3408;
e-mail: ashgriz@mie.utoronto.ca

R   thermal resistance
t   time
$t_{solid.}$ solidification time
$t_{spr.}$ spreading time
$T_{avg.}$ average temperature of droplet
T_d impact temperature of droplet
T_j jetting temperature of printhead
$T_{liq.}$ liquidus temperature
T_s substrate temperature
$T_{sol.}$ solidus temperature of ink
U   droplet velocity
x   one-dimension coordinate

α specific heat
λ superheat parameter
μ viscosity
ρ density
σ surface tension
θ_c calculated contact angle
θ_e equilibrium contact angle
θ_s measured apparent solidification contact angle

Dimensionless numbers

Bo   Bond number
Ca   Capillary number
Oh   Ohnesorge number
Pr   Prandtl number
Re   Reynolds number
St   Stephan number
We   Weber number

1. INTRODUCTION

In a solid inkjet printer, the printhead jets an image first onto an intermediate transfer drum. Once the complete image has been deposited on the transfer drum, the page is brought into contact with the drum through a high-pressure nip and the image is transferred onto paper [1]. Instead of water-based inks, solid inkjet printers use hot-melt inks (wax-based) in that they are in solid state at room temperature. The ink is heated above its melting point and is ejected as droplets onto a drum, the temperature of which is lower than the melting temperature of the ink. Many of the print quality attributes of the solid inkjet printer are largely determined by the intermediate image formed on the drum surface.

In an isothermal droplet impact where the droplet and substrate have the same temperature, the final contact angle is an equilibrium contact angle, which is determined by the droplet material, substrate material and temperature. During the printing-on-drum process as mentioned above, the final contact angle does not represent a three-phase thermal equilibrium state. This final contact angle is referred to as apparent solidification contact angle, which has been the focus of the present work.

There is a large literature base on droplet impaction on solid surfaces, and most of these studies [2–13] have focused on the spreading diameter and splat height. Apparent contact angle was also investigated in some previous works [14–19]. Bhola and Chandra [16] observed the apparent dynamic contact angles of 2 mm single molten paraffin wax droplets impacting on a stainless steel surface with three substrate temperatures, T_s. With the melting temperature of the paraffin wax being 70°C, impaction under $T_s = 40°C$ and 73°C ended up with the same contact angle as the equilibrium contact angle at 73°C, while a larger contact angle was formed at $T_s = 23°C$. Attinger *et al.* [17] recorded the evolution of dynamic contact angle during the deposition of molten solder droplets on solid substrates, and concluded that the contact angle dynamics was strongly coupled to the evolution of the droplet free surface, and no quantitative agreement with Hoffman's law [20] was found. Schiaffino and Sonin [14] studied the apparent dynamic contact angles formed by continuous droplet deposition on a homologous substrate, where the droplets and substrate were the same material. The contact angle was correlated with Stephan and capillary numbers. For homologous droplet impaction with a low Weber number (We ≪ 1) and negligible viscosity effect, Schiaffino and Sonin [15] assumed the final shape of the droplet as a spherical cap and proposed a model to predict the apparent solidification contact angle depending on the Stefan number and materials. Sikalo *et al.* [19] experimentally and numerically investigated the dynamic contact angle of isothermal impaction of single droplets on substrates with different wettabilities.

All these early studies [14–19] show that the evolution of apparent dynamic contact angle can be affected by several factors such as substrate temperature, droplet size and velocity, droplet and substrate materials and so on. It was also found that the apparent dynamic contact angle had significant effect on the impaction dynamics [4, 17, 18]. However, still no systematic study has been done on the variations of apparent solidification contact angle with impact conditions, including substrate temperature, droplet velocity, droplet and substrate materials etc. This work experimentally investigates one aspect of the printing process on the drum surface in solid inkjet printers by considering single droplets of hot-melt ink impacted on solid surfaces. Apparent solidification contact angle, which is called apparent contact angle hereafter, was investigated. Various types of substrates, substrate temperatures, flight distances of droplets, and jetting temperatures were examined.

2. EXPERIMENTAL DETAILS

2.1. Experimental setup

Figure 1 illustrates the experimental setup, which comprises a printhead (Phaser 860 provided by Xerox Corporation) and a substrate. A droplet, initially at T_j, is ejected out of the printhead. After flying a distance L, the droplet impacts a solid surface, the temperature of which is denoted by T_s. Both T_j and T_s were well controlled with a fluctuation range of $\pm 0.5°C$. The droplet temperature upon impact is the impact temperature T_d. Four experimental parameters were varied to change the impact conditions, namely the type of substrate surface, the distance L, the substrate temperature T_s, and the jetting temperature T_j. Three types of substrates were used. One was an uncoated aluminum substrate. The roughness of this uncoated surface was measured to be 0.05 μm (average roughness). The second was a Viton-coated substrate obtained by dip coating the aluminum substrate with a Viton (DuPont, Wilmington, DE) layer 1.8 μm thick. By using Scanning Electron Microscopy (SEM), it was found that there was no significant difference between these two surfaces in terms of roughness. The third type of substrate was made by depositing a silicone oil layer 1 to 2 μm thick on the aluminum substrate surface, which is referred as silicone-oil coated substrate hereafter. The substrates were put at two locations, $L = 0.5$ mm and 1 mm from the printhead. The substrate temperature was in the range from 60°C to 80°C at an interval of 5°C, namely $T_s = 60°C$, 65°C, 70°C, 75°C, 80°C. Two jetting temperatures, $T_j = 140°C$ and 145°C, were examined. As listed in Table 1, four sets of experiments, which are referred to as

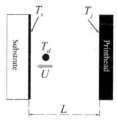

Figure 1. Sketch of the printhead and substrate.

Table 1.
List of four sets of experiments conducted in the present work

Experiment	Substrate	T_j
Case 1	Uncoated	140°C
Case 2	Viton-coated	140°C
Case 3	Silicone-oil coated	140°C
Case 4	Uncoated	145°C

four cases hereafter, were conducted. Each experimental condition specified by the type of substrate, T_s, L and T_j was repeated five times.

The ink used was ColorStix 8200 manufactured by Xerox Corporation. It melts over a range of $T_{sol.} = 60°C$ to $T_{liq.} = 115°C$, and is a non-Newtonian fluid. The temperature relations of viscosity μ and surface tension σ of the ink were measured and are plotted in Fig. 2. It should be noticed that the viscosity of ColorStix 8200 sharply increases at 90°C. The surface tension of ColorStix 8200 increases from 25.54 mN/m at 140°C to 26.45 mN/m at 120°C. This indicates that ColorStix 8200

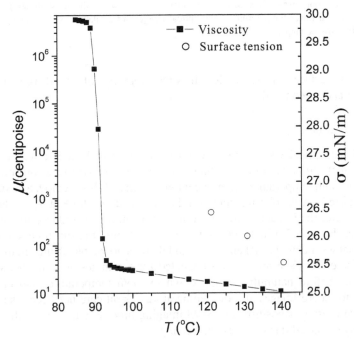

Figure 2. Viscosity and surface tension of Colorstix 8200 versus temperature. The viscosity was measured when the temperature was decreasing.

Table 2.
Properties of the ink, Viton, aluminum and silicone oil used in the experiments

Property	Density (kg/m³)	Specific heat (kJ/(kg·K))	Thermal conductivity (W/(m·K))	Solidus temperature (°C)	Liquidus temperature (°C)
ColorStix 8200	820	2.25	0.18	60	115
Viton	1915	955	0.17	–	–
Aluminum	2770	0.875	177	–	–
Silicone oil	–	–	0.1	–	–

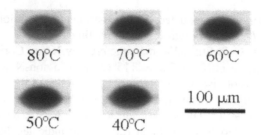

Figure 3. The droplet was deposited at $T_s = 80°C$, and afterwards the substrate temperature was reduced to several lower temperatures. The optical images were taken *in situ*, showing no visible change in the droplet shape.

features low surface tension. Table 2 lists the properties of ColorStix 8200, Viton, aluminum and silicone oil.

2.2. Visualization

Conventional optical visualization methods cannot provide satisfactory resolution for such small droplets. In the present work, SEM was used to record pictures of droplets at room temperature. To prove the feasibility of this method, an experiment was conducted. A droplet impacted on the uncoated substrate at $T_s = 80°C$, and 5 minutes after the impaction the substrate temperature was reduced to a lower T_s. There was always a 5-minute annealing before reducing the substrate temperature. Figure 3 shows the images taken *in situ*, and there is no visible change in the droplet shape. Therefore, each droplet was annealed for 5 minutes at its impact substrate temperature before naturally cooling down to room temperature, and then SEM was used to record images at room temperature. Great care has been taken during conductive coating deposition and SEM imaging processes to prevent the droplets from being damaged due to excessive heating.

2.3. Droplet size, velocities and temperatures

A high-speed video camera was used with a microscope lens to observe the droplets in flight and to measure the droplet size and velocity. The droplet was observed to be spherical at $L = 0.5$ mm. This indicates that the droplet was not in the excited state for the two distances, $L = 0.5$ mm and 1 mm. The droplet diameter, D, was 39 μm. The droplet velocity, U, at $L = 0.5$ mm was 2.81 m/s, and it decreased to 2.56 m/s at $L = 1$ mm.

The in-flight cooling of droplets, which was usually neglected in early studies [3, 14, 16], was considered. When flying from the printhead to substrate surface, the droplet is decelerated due to air drag, and its temperature decreases due to convective heat transfer. The Biot number of the droplet was estimated to be less than 0.1, and the lumped capacitance method [21], therefore, was employed. This

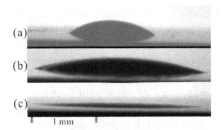

Figure 4. Pictures showing equilibrium shapes of molten ink on three types of substrates at 120°C. (a): silicone oil coated substrate (Bo = 0.1); (b): Viton-coated substrate (Bo = 0.4); (c): uncoated substrate (Bo = 0.1).

heat transfer problem was numerically solved by considering the instantaneous droplet velocity. The results show that the droplet temperature decreases from $T_j = 140°C$ to $T_d = 136.2°C$ at $L = 0.5$ mm and to $T_d = 132.3°C$ at $L = 1$ mm. These two temperatures are the impact temperatures of droplets at the two flight distances. The calculation shows that increasing the jetting temperature to $T_j = 145°C$ causes the two impact temperatures to increase by around 5°C.

3. RESULTS

3.1. Equilibrium contact angles and thermal effects of substrate surfaces

The wettability of substrate has been observed to have effects on the droplet impaction dynamics [2, 9, 18]. Therefore, the apparent contact angle could be affected by the wettability of substrate. In view of this, we first measured the equilibrium contact angle θ_e of the ink on the three types of surfaces.

Due to the nonexistence of three-phase thermal equilibrium at the droplet-substrate contact line, no equilibrium contact angles exist under the experimental conditions to be presented in this paper. Hence, θ_e was measured at 120°C, 5°C higher than the liquidus temperature of the ink. Small lumps of solid ink were chosen and weighed. The mass of ink lump, m, was used to calculate the volume of molten droplet and its Bond number. The Bond number compares the gravitation with the surface tension and is defined as

$$\text{Bo} = \left(\frac{36m^2\rho}{\pi^2} \right)^{1/3} \frac{g}{\sigma} \tag{1}$$

The ink lumps were placed on the substrate surface when T_s was 75°C, and then T_s was increased to 110°C. At 110°C, the ink quickly melted and spread. After the ink reached a stationary state, T_s was slowly increased to 120°C. Figure 4 shows three molten droplets in equilibrium states on the three substrates.

Several lumps of ink with different masses were used, and θ_e was estimated by linearly fitting the measurements to zero Bond number. With this method, θ_e was found to be 28° for the silicone-oil coated surface, and 12° for the Viton-coated surface. The ink on the uncoated surface spread into a very thin flat layer. The contact angle was between 0.8° and 1.1° and did not vary with the melt volume significantly. Hence, simply averaging all the measurements, θ_e on the uncoated surface was estimated to be 1°.

Since the thermal conductivity of aluminum is much higher than that of Viton and silicone oil (see Table 2), here we compare the thermal resistance of Viton and silicone oil layers with that of the ink droplet, as:

$$\frac{R}{R_d} = \frac{lk_d}{Dk} \tag{2}$$

where R_d is the thermal resistance of droplet, and R, k and l are the thermal resistance, conductivity and thickness of Viton or silicone oil layer. This thermal resistance ratio is 0.05 for the Viton coated substrate, and 0.09 for the silicone oil coated substrate. Therefore, the thermal resistances of Viton layer and silicone oil layer are negligible.

3.2. Droplets impacted on substrate surfaces

Figures 5 through 8 present the final shapes of droplets impacted on the three types of substrates under different impact conditions. It can be seen that not only the droplet shape but also the surface texture varies with the type of substrate,

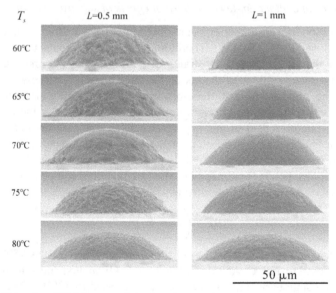

Figure 5. Side view SEM images of droplets impacted on the uncoated aluminum substrate, $T_j = 140°C$ (Case 1).

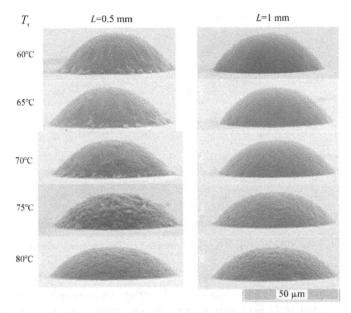

Figure 6. Side view SEM images of droplets impacted on the Viton-coated substrate, $T_j = 140°C$ (Case 2).

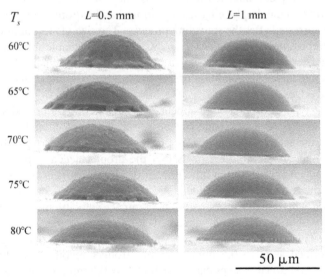

Figure 7. Side view SEM images of droplets impacted on the silicone-oil coated substrate, $T_j = 140°C$ (Case 3).

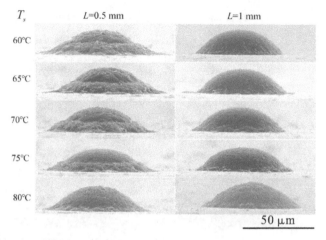

Figure 8. Side view SEM images of droplets impacted on the uncoated substrate, $T_j = 145°C$ (Case 4).

substrate temperature, flight distance and jetting temperature. Generally, the droplets impacted at $L = 1$ mm assume regular shapes and smooth textures as compared with those at $L = 0.5$ mm.

Figure 5 shows the droplets impacted on the uncoated aluminum substrate with $T_j = 140°C$. Droplets impacted at low substrate temperatures and $L = 0.5$ mm have thin skirts on the droplet edges, suggesting that the droplets recoiled with their contact lines arrested after the spreading process. The size of skirt diminishes as the substrate temperature increases. At $T_s = 80°C$, no appreciable difference between the two distances is observed. As shown in Fig. 6, droplets impacted on the Viton-coated substrate under the condition of low substrate temperatures and $L = 0.5$ mm show jagged edges.

A different type of droplet edge appears on the silicone-oil coated substrate, as shown in Fig. 7. A jagged edge can be see at $T_s = 60°C$ and $L = 0.5$ mm, but as substrate temperature increases one can see "concave steps" formed near the contact line. A possible reason is that the droplet needs to push away some silicone oil when spreading, and the impaction dynamics near the contact line is affected by the resistance from the silicone oil layer. This effect was significant at $L = 0.5$ mm where the droplets had higher impact velocity and temperature than those at $L = 1$ mm. Using the uncoated aluminum substrate but higher jetting temperature $T_j = 145°C$, Fig. 8 shows very different pictures than Fig. 5. Larger skirts are formed at $L = 0.5$ mm, showing stronger recoil of the droplets with $T_j = 145°C$. Additionally, steps can be seen at $L = 0.5$ mm, which become smaller as the substrate temperature decreases.

A common observation made from Figs. 5–8 is that droplets impacted at $L = 0.5$ mm show rough surface textures. This could be due to the droplet crystallization, which is affected by the flow and thermal histories of droplets during impact.

3.3. Apparent solidification contact angles

Before presenting the measurements of apparent contact angle, we conduct an analysis, which will contribute to the discussion later. In the present work, when a droplet starts spreading on a sub-cooled surface, a portion of the kinetic energy is consumed by viscous dissipation, and the remaining kinetic energy flattens the droplet, causing the increase of surface energy. Strictly speaking, there is never complete solidification during the impaction process, since all the substrate temperatures employed are within the mushy zone (between solidus temperature 60°C and liquidus temperature 115°C) of the ink. However, both viscosity and surface tension increase during the impaction process, due to the temperature dependence of both properties (see Fig. 2). Bennett and Poulikakos [2] concluded that surface tension effect dominates the termination of droplet spreading over viscous effect when

$$\text{We} \ll 2.8\text{Re}^{0.457} \tag{3}$$

where $\text{We} = \rho U^2 D/\sigma$ and $\text{Re} = \rho U D/\mu$. In the present work, when $T_j = 140°C$, Weber numbers are 8.8 and 7.3 at $L = 0.5$ mm and 1 mm, respectively, and the corresponding Reynolds numbers are 7.7 and 6.4. Hence, the corresponding values on the right hand side are 7.5 and 7.2, respectively. Therefore, the viscous dissipation has a significant effect.

A more substantial support for this conclusion can be found by evaluating a dimensionless number,

$$\text{Ca} = \frac{\mu U}{\sigma} \tag{4}$$

Ca is called the capillary number, in which the velocity of the moving contact line is approximated as the droplet impact velocity. The variations of viscosity and surface tension with temperature as shown in Fig. 2 were used to evaluate the change in capillary number with temperature. Since the surface tension was measured only at three temperature points, linear interpolation and extrapolation were applied to estimate surface tension at other temperature points. The results are presented in Fig. 9, where one can see that the capillary number increases linearly from 1.2 at 140°C to 3.9 at 94°C and sharply increases afterwards. This indicates that surface tension has only a minor effect at the beginning of droplet impact, and that the domination of viscous effect increases as the droplet impact proceeds. Therefore, the droplet impaction is a coupling process of viscous damping and impact inertia. More discussion will be given after presenting the measured apparent contact angle.

A drop-shape analysis software (DropSnake and LB-ADSA, The National Institutes of Health, Bethesda, MD) was used to determine the apparent contact angle by finding the local tangent angle to the droplet edge. The method is illustrated in Fig. 10. Two angles, θ_l and θ_r, were measured, and the average of which was considered as the contact angle, θ_s.

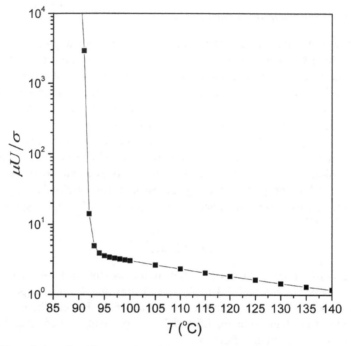

Figure 9. The variation of capillary number with temperature ($U = 2.81$ m/s).

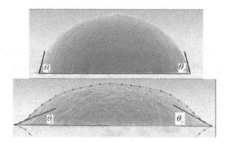

Figure 10. The method of measuring the solidification contact angle. θ_l and θ_r are first measured, and then the average of the two angles is θ_s.

The measurement results are presented for the four cases as shown in Table 1. To clearly show the tendencies, simple linear fittings were conducted with respect to substrate temperature. Figure 11 compares the uncoated aluminum substrate with the Viton-coated substrate with $T_j = 140°C$ (Cases 1 and 2). The apparent contact angle decreases with increasing substrate temperature. Apparent contact angles formed at $L = 1$ mm are larger than those formed at $L = 0.5$ mm. This is because the impact velocity and temperature at $L = 0.5$ mm are higher than

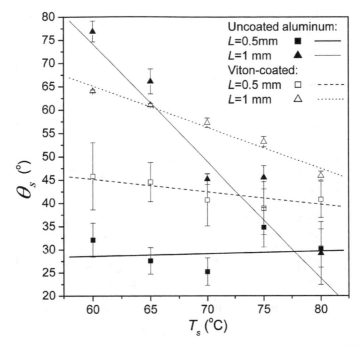

Figure 11. Apparent contact angles formed on the uncoated aluminum substrate and Viton-coated substrate, $T_j = 140°C$ (Cases 1 and 2).

those at $L = 1$ mm. As discussed above, the droplet impaction in the present work is a coupling process of viscous damping and inertia-driven impact. We compare this coupling effect at the two distances by comparing the Reynolds numbers of the droplet at these distances, as:

$$\frac{\text{Re}_{L=0.5}}{\text{Re}_{L=1}} = \frac{U_{L=0.5}\mu_{L=1}}{U_{L=1}\mu_{L=0.5}} \tag{5}$$

Interpolating from Fig. 2, viscosity of droplet is 11.5 cP (centipoise) at $L = 0.5$ mm ($T_d = 136.2°C$) and 12.7 cP at $L = 1$ mm ($T_d = 132.3°C$). The Reynolds number ratio is around 1.22, which indicates that the viscous damping has more effect at $L = 1$ mm. Hence, larger apparent contact angles are formed $L = 1$ mm due to early arrest of contact line or less recoiling.

Figure 11 also shows that the slopes at $L = 1$ mm are steeper than those at $L = 0.5$ mm for both substrates, showing a higher sensitivity to the substrate temperature at $L = 1$ mm. As the substrate temperature rises, the effect of viscous damping decreases. Supposing that the decrease of substrate temperature causes the average viscosity of the impacting droplet to increase by $\Delta\mu$, then the effect of this change can be evaluated by $\Delta\mu/\rho U D$. Since the impact velocity U is

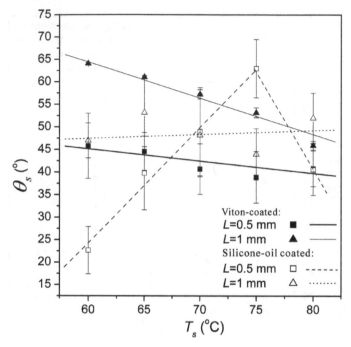

Figure 12. Apparent contact angles formed on the Viton-coated substrate and silicone oil coated substrate, $T_j = 140°C$ (Cases 2 and 3).

lower at $L = 1$ mm, the value of $\Delta\mu/\rho U D$ is larger at $L = 1$ mm than that at $L = 0.5$ mm. This implies that the droplet impaction at $L = 1$ mm is more sensitive to the variation of substrate temperature.

Generally, the contact angles formed on the uncoated aluminum substrate are smaller than those on the Viton-coated substrate, showing the effect of wettability. However, exceptions can be found at $T_s = 60°C$ and $65°C$ with $L = 1$ mm, where larger contact angles are formed on the uncoated aluminum substrate. This is similar to the phenomenon observed in Schiaffino and Sonin [15] that early contact line arrest by freezing can induce behavior similar to the effect of a large equilibrium contact angle. For both substrates, contact angles formed at these two distances tend to be equal as the substrate temperature increases. This is due to the low viscous effect at high substrate temperature.

Figure 12 shows the comparison between the Viton-coated substrate and the silicone-oil coated substrate (Cases 2 and 3). One can see that the comparison does not show the effect of equilibrium contact angle, which has been observed in the comparison of the Viton-coated substrate with the uncoated substrate in Fig. 11. This is because the existence of a liquid layer of silicone oil on the solid surface affects not only the wettability but also the fluid dynamics near the contact line.

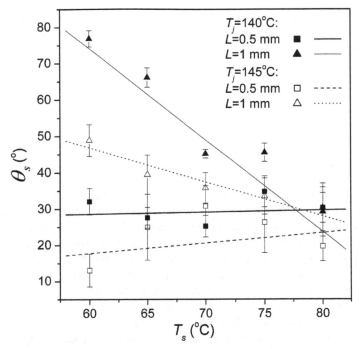

Figure 13. Apparent contact angles formed on the uncoated aluminum substrate with two jetting temperatures $T_j = 140°$C and $145°$C (Cases 1 and 4).

A portion of the kinetic energy of the droplet is consumed on the droplet edge to push away some silicone oil. This is evident in Fig. 7, where the effect of the silicone oil layer can be seen on the droplet edges at $L = 0.5$ mm. At $L = 1$ mm, due to a lower impact velocity, the apparent contact angles are mainly affected by the resistance of silicone oil and the viscous damping inside the droplets, remaining almost constant for all substrate temperatures. However, the apparent contact angles formed at $L = 0.5$ mm are found to increase with substrate temperature from 60°C to 75°C, but decrease from 75°C to 80°C (see Fig. 12). At low substrate temperatures, the apparent contact angles are due to the coupling effect of spreading and viscous damping. As the substrate temperature increases, the droplet tends to spread more, and the resistance from the silicone oil tends to prevent the contact line from further spreading, thereby causing the apparent contact angle to increase. The decrease of apparent contact angle from $T_s = 75°$C to 80°C is mainly due the weak effect of viscous damping at high substrate temperature and also due to the decrease of viscosity of silicone oil.

Figure 13 shows the apparent contact angles formed on the uncoated aluminum substrates with two jetting temperatures: $T_j = 140°$C and $145°$C (Cases 1 and 4). First, the apparent contact angles formed with the higher jetting temperature are smaller than those formed with lower jetting temperature. As mentioned above, the droplet impact temperatures in the case of $T_j = 145°C$ are five degrees higher than those in the case of $T_j = 140°C$, and the viscous effect lessens as the droplet temperature increases. Second, the apparent contact angles formed at $L = 1$ mm are less dependent on the substrate temperature in the case of $T_j = 145°C$ than in the case of $T_j = 140°C$. The effect of varying the jetting temperature dwindles as the substrate temperature increases to 80°C.

4. DISCUSSION

The experimental results presented above show that the apparent contact angles vary with the type of substrate and substrate temperature, which are also the major factors in affecting the equilibrium contact angles. However, the results also show the significant effect of droplet flight distance, which changes the impact velocity and temperature of the droplet, as well as the jetting temperature. All these factors considered in the present work affect the fluid behavior near the contact line of the droplet during impact, which, in turn, affects the evolution of the droplet free surface. The evolution of free surface near the contact line eventually manifests itself in the formation of solidification contact angle.

Gao and Sonin [3] concluded that a solidified droplet possesses an apparent static contact angle which is a property of the molten material, the substrate material and the substrate temperature, but is independent of the spreading process. In the other words, the droplet impaction could end up with a spherical sessile droplet due to a mechanical but not a thermal equilibrium shape, if the solidification time was much longer than the impaction time. Therefore, apparent contact angles could be calculated using the geometric relations for a spherical cap. Before evaluating the applicability of this method to the present work, we will first conduct an analysis on the time scales of droplet impact and solidification.

Gao and Sonin [3] proposed a simple model to estimate the solidification time scale of a molten droplet impacting on a cold solid surface, which can be rearranged as

$$t_{solid.} \approx \frac{2D^2}{3\alpha} \frac{k_d}{k_s} \left[\ln(\lambda + 1) + St^{-1} \right] + \frac{D^2}{3\alpha} St^{-1} \qquad (6)$$

Here λ and St are the superheat parameter and Stefan number, respectively. The expressions for these two parameters are

$$\lambda = \frac{T_d - T_{liq.}}{T_{liq.} - T_s}, \qquad St = \frac{C_p(T_{liq.} - T_{sol.})}{L_f}$$

In the present case, the term $\ln(\lambda + 1)$ ranges from 0.27 to 0.47, and St is 0.676. The latent heat, L_f, has been assumed to liberate uniformly over the melting range of 60°C to 115°C. Besides, the ratio of thermal conductivities is of the order of 10^{-3}.

Hence, Eq. (6) can be simplified as

$$t_{solid.} \approx \frac{D^2}{3\alpha} \text{St}^{-1} \tag{7}$$

In the present work, The Ohnesorge number, $\text{Oh} = \mu(\rho D\sigma)^{-0.5}$, is 0.36. According to Schiaffino and Sonin [15], the characteristic spreading time $t_{spr.}$ under the condition of $\text{We} \gg 1$ and $\text{Oh} \ll \text{We}^{0.5}$ can be approximately estimated by

$$t_{spr.} = \frac{D}{U} \tag{8}$$

Hence, combining Eqs. (7) and (8), the ratio of the spreading and solidification time scales in the present work can be estimated by

$$\frac{t_{spr.}}{t_{solid.}} \approx 3 \frac{\text{St} \cdot \text{Oh} \cdot \text{We}^{0.5}}{\text{Pr}} \tag{9}$$

where Pr is the Prandtl number, which is expressed by $\text{Pr} = \mu C_p/k_d$. The calculation results are 0.0055 and 0.005 for $L = 1$ mm and 0.5 mm, respectively. This indicates that the spread time is 3 orders of magnitude lower than the solidification time. This result seems to justify using the following geometric relations to calculate the apparent contact angles for our case.

$$\text{If } \frac{D^3}{h^3} \leqslant 4, \quad \theta_c = \pi - \sin^{-1}\left(\frac{\sqrt{12D^3h^3 - 12h^6}}{2h^3 + D^3}\right)$$

$$\text{If } \frac{D^3}{h^3} \geqslant 4, \quad \theta_c = \sin^{-1}\left(\frac{\sqrt{12D^3h^3 - 12h^6}}{2h^3 + D^3}\right) \tag{10}$$

where h is the height of the sessile droplet, and θ_c is the apparent angle obtained via this method. The first relation is for blunt angles, while the second is for sharp angles.

Only Cases 1 and 2 are considered here. The heights of the droplets impacted on the uncoated aluminum and Viton-coated substrates ($T_j = 140°C$) are measured, which are then used to calculate the angles θ_c. The difference between the calculated and measured angles, $\theta_s - \theta_c$, is plotted in Fig. 14, and dramatic differences can be found. For instance, under the condition of $L = 0.5$ mm and $T_s = 60°C$ on the uncoated surface θ_s is 35° smaller than θ_c, while under the condition of $L = 1$ mm on the Viton-coated surface θ_s is very close to θ_c. Several observations can be made from Fig. 14. First, θ_s is generally smaller than θ_c. Second, for both distances, larger differences are observed on the uncoated aluminum substrate; for both substrates, larger differences occur at $L = 0.5$ mm. Third, as T_s increases, the difference decreases for the case of $L = 0.5$ mm, but increases in the case of $L = 1$ mm. Fourth, as T_s increases, the difference caused by the flight distances L diminishes.

All these observations from Fig. 14 show that the final shape of the droplet deviates from the shape of a spherical cap to varying degrees depending on the impact condition. In the time scale analysis above, the bulk solidification of the droplet is considered. Since all the substrate temperatures employed are less than 90°C where the ink is highly viscous, the contact line is arrested due to the formation of a highly viscous layer at the droplet-substrate interface. Upon the arrest of contact line, the remaining part of the droplet would retract, thereby causing smaller contact angles that those determined using Eq. (10). No information is available for a detailed discussion due to the difficulty in visualizing the dynamic impaction of such small droplets. Below a temperature analysis to estimate the formation of the highly viscous layer in the droplet is shown.

A thermal resistance analysis shows that the internal resistance of an ink droplet is three orders of magnitude higher than that of the substrate and more than one order of magnitude lower than the air. Therefore, we consider a one-dimensional heat conduction problem in a finite slab with an insulated boundary and a constant temperature boundary. Simplifying the method in Bulavin and Kashcheev [22] to

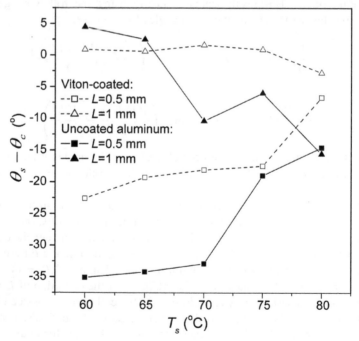

Figure 14. The difference between the calculated contact angle and the measured contact angle: $\theta_s - \theta_c$.

one slab, the solution to this problem is

$$\frac{T\,(x,t) - T_s}{T_d - T_s} = \sum_{n=1}^{\infty} \frac{4}{(2n-1)\,\pi} \exp\left(-\frac{(2n-1)^2\,\pi^2}{4}\frac{\alpha t}{h^2}\right) \sin\left[\frac{(2n-1)\,\pi}{2}\frac{x}{h}\right]$$

(11)

where h is the thickness of the slab, and x is the distance from the ink-substrate interface. The thickness of the slab h is approximated as 20 μm, the average of droplet heights in the present work.

By using Eq. (11), we can find two temperature fronts: viscous front ($T = 90°$C) and mushy front ($T = 115°$C) at maximum spreading time scales given by Eq. (8). Here we only consider $T_j = 140°$C and neglect the thermal effect of coating layer. Figure 15 shows that the thickness of the two layers for different substrate temperatures T_s and distances L. Both layers are thicker at $L = 1$ mm, and their thickness decreases as T_s increases. This corresponds to the analysis of viscous effect mentioned above. As mentioned in Fig. 5, some skirts are observed on the droplet edges in the case of low substrate temperatures and $L = 0.5$ mm. In Fig. 16, the droplets impacted at $T_s = 60°$C and 65°C with $L = 0.5$ mm in Fig. 5 are enlarged to show the details on the edges. The skirt thicknesses were measured to be 1.6 μm and 1.2 μm (see Fig. 16), which are in between the mushy and viscous

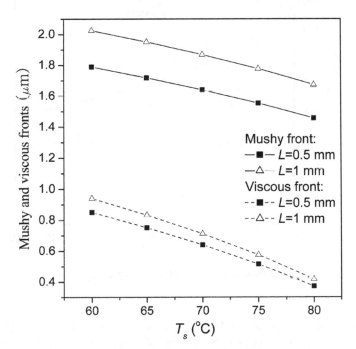

Figure 15. Thicknesses of mushy and viscous layers formed at the time of maximum spreading.

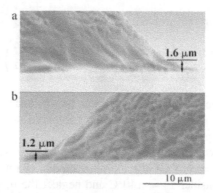

Figure 16. Thicknesses of skirts at the droplet edges. (a: $T_s = 60°C$; b: $T_s = 65°C$; $L = 0.5$ mm; $T_j = 140°C$; the uncoated aluminum substrate).

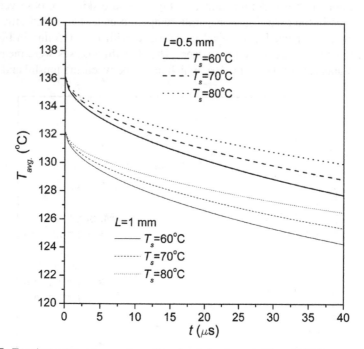

Figure 17. Transient average temperature of droplet during impact. ($T_j = 140°C$).

layer thicknesses as predicted in Fig. 15. Due to the higher impact temperature and velocity at $L = 1$ mm, droplets may recoil after spreading. When recoiling, the fluid in the mushy and viscous layers (see Fig. 15) tends to remain and the rest of the droplet tends to withdraw, thereby forming skirts around the droplet edge.

An expression for the transient average temperature of droplet can also be obtained from Eq. (11) as:

$$\frac{T_{avg.}\,(t) - T_s}{T_d - T_s} = \sum_{n=1}^{\infty} \frac{8}{(2n-1)^2\,\pi^2} \exp\left(-\frac{(2n-1)^2\,\pi^2}{4}\frac{\alpha t}{h^2}\right) \tag{12}$$

Figure 17 plots Eq. (12) at three substrate temperatures for a time period of 40 μs. First, the droplet average temperature is lower at $L = 1$ mm than at $L = 0.5$ mm; second, droplets impacting at lower substrate temperatures have lower average temperatures. This supports the analysis of the viscous effect in the previous section.

5. CONCLUSIONS

The deposition of ink droplets in a solid inkjet printer was studied by considering single droplets impacting on sub-cooled solid surfaces. This work experimentally explores a range of ink-jet printing conditions by varying four parameters: the type of substrate, substrate temperature, distance between the substrate surface and printhead, and the printhead jetting temperature. Although the theoretical analysis shows that the solidification time scale is much longer than the impaction time scale, the apparent contact angle shows a strong dependence on the impact velocity and temperature of the droplet. This indicates a significant effect of flow history on the formation of apparent contact angle, and this effect lessens at high substrate temperatures. Smaller apparent contact angles are formed on the uncoated aluminum substrate than on the Viton-coated substrate, due to a better wettability of the aluminum surface. However, although the silicone-oil coated substrate shows the highest equilibrium contact angle, the existence of a silicone oil layer affects the fluid dynamics of the droplet edge, due to coupling of two fluids but not due to difference in wettability. It is also found that the deposited droplet deviates from the shape of a spherical cap to varying degrees depending on the impact conditions, and assuming a spherical cap could cause significant error in apparent contact angle.

Acknowledgements

The authors are indebted to Dr. Farhad Farhadi, who is involved in the numerical simulation of this project, for his continuous and valuable input to this project. The authors are also indebted to Chris Wagner and Bradley Gerner for their assistance in setting up the jetting fixture and to David Gervasi for the Viton coating. This work is supported by the Xerox Foundation and the Natural Sciences and Engineering Research Council of Canada (NSERC).

REFERENCES

1. T. Snyder and S. Korol, in: *Recent Progress in Ink Jet Technologies II*, The Society for Imaging Science and Technology, pp. 175-181, Springfield, Virginia (1999).

2. T. Bennett and D. Poulikakos, *J. Mater. Sci.* **28**, 963 (1993).
3. F. Gao and A. A. Sonin, *Proc. R. Soc. Lond. A* **444**, 533 (1994).
4. J. Fukai, Y. Shiiba, T. Yamamoto and O. Miyatake, *Phys. Fluids* **7**, 236 (1995).
5. Z. Zhao, D. Poulikakos and J. Fukai, *Int. J. Heat Mass Transfer* **39**, 2791 (1996).
6. J. M. Waldvogel and D. Poulikakos, *Int. J. Heat Mass Transfer* **40**, 295 (1997).
7. S. Schiaffino and A. A. Sonin, *Phys. Fluids* **9**, 2227 (1997).
8. S. D. Aziz and S. Chandra, *Int. J. Heat Mass Transfer* **43**, 2841 (2000).
9. H.-Y. Kim and J.-H. Chun, *Phys. Fluids* **13**, 643 (2001).
10. S. Haferl and D. Poulikakos, *J. Appl. Phys.* **92**, 1675 (2002).
11. R. Rioboo, M. Marengo and C. Tropea, *Experiments in Fluids* **33**, 112 (2002).
12. D. B. Dam and C. L. Clerc, *Phys. Fluids* **16**, 3403 (2004).
13. S.-C. Gong, *Jap. J. Appl. Phys.* **44**, 3323 (2005).
14. S. Schiaffino and A. A. Sonin, *Phys. Fluids* **9**, 2217 (1997).
15. S. Schiaffino and A. A. Sonin, *Phys. Fluids* **9**, 3172 (1997).
16. R. Bhola and S. Chandra, *J. Mater. Sci.* **34**, 4883 (1999).
17. D. Attinger, Z. Zhao and D. Poulikakos, *J. Heat Transfer – Trans. ASME* **122**, 544 (2000).
18. A. Amada, M. Haruyama, T. Ohyagi and K. Tomoyasu, *Surface Coatings Technol.* **138**, 211 (2001).
19. S. Sikalo, H.–D. Wilhelm, I. V. Roisman, S. Jakirlic and C. Tropea, *Phys. Fluids* **17**, 062103 (2005).
20. R. L. Hoffman, *J. Colloid Interface Sci.* **50**, 228 (1975).
21. F. P. Incropera and D. P. Dewitt, *Fundamentals of Heat and Mass Transfer*, 3rd ed., pp. 226-229, John Wiley & Sons, New York (1990).
22. P. E. Bulavin and V. M. Kashcheev, *Intl. Chem. Eng.* **5**, 112 (1965).

Contact Angle, Wettability and Adhesion, Vol. 5, pp. 47–57
Ed. K.L. Mittal
© VSP 2008

Microscopic wettability of softwood tracheid cell walls: Effects of wall sculpturing, sap solutes, and drying

MIKA M. KOHONEN*

Department of Applied Mathematics, Research School of Physical Sciences and Engineering, Australian National University, Canberra ACT 0200, Australia

Abstract—The wettability of the inner (lumen) walls of the water-conducting capillaries in plants (tracheids and vessels) is an important parameter in both plant physiology and wood technology. However, although the macroscopic wettability of wood surfaces has received considerable attention, lumen wettability has not been studied in detail. Here measurements of the wettability of the lumen walls of tracheids of two species of Australian cypress (*Callitris* Vent.) are described. It is shown that wall roughness (the warty layer) significantly increases the wettability of the lumen wall, an observation which provides an appealing answer to the long-standing debate on the function of wall sculpturing in plant capillaries, and may provide clues for the biomimetic engineering of wettability in technological applications such as microfluidics. The results also demonstrate that lumen wettability is enhanced by the presence of hygroscopic sap residues, and is diminished by air-drying (aging).

Keywords: Contact angle; wettability; xylem; plants; embolism; wood; water transport; tracheid.

1. INTRODUCTION

Water moves from the roots to the leaves of vascular plants through a network of capillaries formed from dead cells, referred to as tracheids and vessels [1, 2]. The wettability of the lumen walls of plant capillaries is thus an important parameter in understanding the transport of water in living plants [1, 2], particularly in models of embolism repair (the removal of flow-blocking bubbles arising from the cavitation of metastable water during transpiration) [3–7]. The contact angle (θ) of water on the lumen walls of plant capillaries also affects the use of wood as a material, influencing, for example, the impregnation of wood with aqueous solutions [8, 9]. The extent to which wood will absorb water, particularly at high humidities, is also affected by lumen wettability

Surprisingly only little is known about the wettability of plant capillaries. Plant physiologists commonly assume that lumen walls are perfectly wetted by water

*Current address: Surfactant and Colloid Group, Department of Chemistry, The University of Hull, Cottingham Road, Kingston upon Hull, HU6 7RX, United Kingdom. Tel.: +44 (0)1482 465283; Fax: +44 (0)1482 466410; e-mail: m.kohonen@hull.ac.uk

($\theta = 0°$). In contrast, opponents of the cohesion-tension theory of sap transport assert that lumen walls are hydrophobic, due to the presence of lignin in the cell wall [10, 11]. There is a large literature on the wettability of macroscopic wood surfaces and the external surfaces of wood fibres (see, for example [12, 13], and references therein), but the results of such studies are difficult to extrapolate to lumen wettability. For example, the external surfaces of fibres may differ in composition from the lumen wall, and may undergo alterations in surface chemistry as a result of the processes used to separate individual fibres from intact wood. The only direct, quantitative study of lumen wettability (of which the author is aware) is that of Zwieniecki and Holbrook [5], who measured the contact angles of water on the lumen walls of six different tree species and reported values of θ between 42 and 55° [5].

There is clearly a need for further studies of lumen wettability, and one might reasonably expect lumen wettability to depend on a range of parameters not explored by Zwieniecki and Holbrook. In this paper, observations on the wettability of the tracheid walls of two species of the Australasian softwood genus *Callitris* Vent. (commonly known as Cypress pine or Australian cypress) are reported. The study was originally motivated by a desire to understand the potential effects of wall sculpturing on lumen wettability. Anatomical studies have revealed a correlation between habitat and the presence of wall sculpturing, or roughness, on the internal surfaces of the capillaries, with sculpturing of various forms (e.g. warty and vestured layers, and helical thickening) being more frequent in species

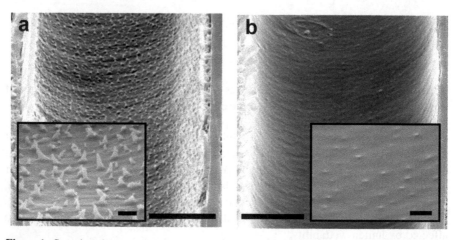

Figure 1. Scanning electron micrographs of the internal surfaces of the water-conducting capillaries (tracheids) in two different species of *Callitris* (Cypress pine, or Australian cypress). (a) The walls of the tracheids in *C. endlicheri* are rough, being covered with a prominent layer of warts, as shown at higher magnification in the inset. (b) The walls of the tracheids in *C. macleayana* are relatively smooth, displaying only small rounded warts, as shown at higher magnification in the inset. Scale bars: (a) and (b), 10 μm; insets, 1 μm.

subject to drought, high transpirational demands, or freezing [14, 15]. This has led to speculation [14] concerning the possible function of roughness in preventing, or reducing, the disruption of water transport which results from blockage of capillaries by air bubbles, or emboli [1, 2], but a satisfactory hypothesis has yet to be established.

Heady *et al.* [15] demonstrated that the tracheid walls in *Callitris* species native to dry habitats are lined with large warts, giving rise to rough capillary walls (Fig. 1a), whilst the warts in species native to high rainfall environments are small and squat, leading to relatively smooth capillary walls (Fig. 1b). Here, the results of measurements on one species with smooth-walled tracheids (*C. macleayana*) and one species with rough-walled tracheids (*C. endlicheri*) are reported. Wettability was studied by analysing microscope images of bubbles produced in the tracheids upon re-wetting of sapwood sections, an approach which has previously been used to study the collapse of emboli [3].

2. MATERIALS AND METHODS

Branches (5–25 mm in diameter) were collected from mature healthy trees growing at the Australian National Botanic Gardens, or on the Australian National University campus, and transported to the laboratory in plastic bags. Thin sections (approximately 1 cm long, 0.5 cm wide and 100–200 μm thick) of sapwood were cut parallel to the length of the tracheids (tangential longitudinal sections) using a clean razor blade. The tracheids in the samples studied were 0.5–2 mm long with a radius of $\sim$ 10 μm, ensuring that most of the tracheids in the prepared sections were complete. Six types of sections were studied: 1. Fresh sections, obtained by cutting sections from fresh sapwood and allowing them to dehydrate in laboratory air for 5–20 minutes; 2. Rinsed-fresh sections, obtained by rinsing fresh sections in a large excess of pure water at room temperature ($\approx$ 20°C) for 1–2 days; 3. Dry-wood sections, obtained by cutting sections from branches which had been dried to constant mass at room temperature and humidity ($\approx$ 50%); 4. Dried-fresh sections, obtained by cutting sections from fresh sapwood, and then drying the sections to constant mass at room temperature and humidity; 5. Rinsed-dry-wood, obtained by rinsing dry-wood sections in a large excess of pure water at room temperature for 1–2 days; 6. Dried-rinsed-fresh sections, obtained by drying rinsed-fresh sections in laboratory air for at least 24 hours.

Wettability was studied by examining air bubbles formed in tracheids upon re-wetting of the sapwood sections. Sections were placed on a clean microscope slide and flooded with pure, filtered water. Sufficient water was added such that when a coverslip was placed on the sections they remained immersed in a continuous film of water. Some measurements were also performed using saturated salt solutions (of $BaCl_2$, NaBr, $NaNO_2$, and K_2CO_3) in place of water. Images of the bubbles were captured approximately 15 minutes after re-wetting, using a digital camera attached to an optical microscope. Sets of between 6 and 10 sections were examined, and

rough walls smooth walls

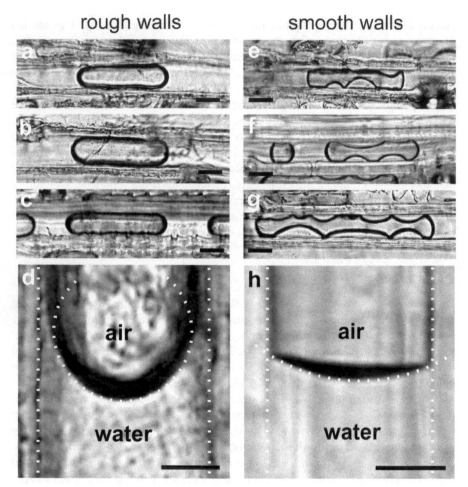

Figure 2. Optical micrographs of air bubbles in tracheids of rehydrated fresh sapwood sections of *C. endlicheri* (a–d) and *C. macleayana* (e–h). The contact angle of water on the lumen walls was obtained by fitting a circle to the air/water meniscus and calculating the angle of intersection with the tracheid walls. Scale bars: (a–c) and (e–g), 20 μm; (d) and (h), 10 μm.

5 images were taken at random from each section. For a given air/water meniscus the contact angle of water on the tracheid wall was obtained by fitting a circle to the meniscus and calculating the angle of intersection with a straight line fitted to the tracheid wall (Fig. 2d, h). When the fitted circle does not intersect the tracheid wall (Fig. 2d) a zero value for the contact angle is recorded. Between 200 and 500 values of contact angle were obtained in this way for a given set of sections. Multiple branch segments were examined for each species, and the results for a

given type of section were obtained from measurements on between 4 and 13 sets of sections.

Finally, some macroscopic measurements of the spreading behaviour of sap on a polymer film were also performed, using a commercial contact angle goniometer (KSV CAM 200). Sap was obtained by squeezing sapwood samples in a vise and collecting the exudate, which was subsequently filtered. A smooth polymer film was prepared by spin-coating Norland Adhesive 73 (NOA73) onto a silicon wafer, followed by uv-curing.

3. RESULTS AND DISCUSSION

First, results of measurements on fresh sapwood sections are discussed. Optical micrographs of typical air bubbles in fresh sapwood sections of *C. endlicheri* and *C. macleayana* are shown in Fig. 2. A clear difference in wettability between the two species was observed. Bubbles in tracheids with a prominent warty layer (*C. endlicheri*) were typically simple in shape (Fig. 2a, b, c), exhibiting a zero contact angle of water on the wall (Fig. 2d). In contrast, the majority of bubbles in smooth-walled tracheids (*C. macleayana*) exhibited a contact angle significantly greater than zero (Fig. 2h) and were irregular in shape (Fig. 2e, f, g), due to pinning of the three-phase contact line, and the presence of droplets on the walls inside the bubbles. Typical histograms of measured contact angles for fresh sapwood sections of *C. endlicheri* and *C. macleayana* are shown in Fig. 3. Approximately 90% of the measured contact angles are zero in tracheids with rough walls (*C. endlicheri*), as compared to only about 40% in tracheids with smooth walls (*C. macleayana*). The spread of contact angles observed in measurements on *C. macleayana* can be attributed to hysteresis of the contact angle, arising from heterogeneity of the lumen wall. Movement of the bubbles in the initial stages of re-wetting will produce a range of contact angles between the receding (θ_R) and the advancing (θ_A) contact angles. The shape of the distribution will depend on details of the re-wetting process, and it is plausible that zero contact angles will sometimes arise as a result of the metastability of thin films produced by receding menisci. Here, the advancing and receding contact angles were, respectively, estimated as the mean plus and minus twice the standard deviation of the distribution of nonzero contact angles. The intrinsic, or equilibrium, contact angle θ_E was then calculated using the equation $\cos \theta_E = 0.5(\cos \theta_A + \cos \theta_R)$ [17]. For *C. macleayana*, $\theta_E = 40 \pm 1°$ (mean ± standard error). The contact angle of the small fraction of nonwetting menisci in *C. endlicheri* was $\theta_E = 43 \pm 3°$. Apparently the intrinsic contact angle of water on the material of which the tracheid walls are composed is approximately 40° for both species, a value which is close to those reported by Zwieniecki and Holbrook [5].

The fact that tracheids in *C. endlicheri* are significantly more wettable than tracheids in *C. macleayana*, a trend which persists across all six types of samples examined (Fig. 4), can be attributed to the presence of a well-developed warty

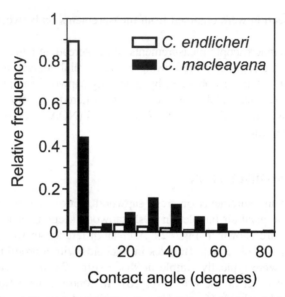

Figure 3. Histograms of contact angles of water on the lumen walls of tracheids in fresh sapwood sections of *C. endlicheri* (211 measurements) and *C. macleayana* (306 measurements).

layer in *C. endlicheri*, as discussed in detail in [16]. Surface roughness amplifies the intrinsic wettability of surfaces [18], and a detailed analysis shows that the lumen walls in *C. endlicheri* are sufficiently rough to reduce the contact angle to near zero, but not significantly rougher [16]. This suggests that wall sculpturing may be a functional adaptation designed to increase the wettability of lumen walls, a phenomenon which is analogous to the well-known Lotus effect [18–20]. This hypothesis may provide an answer to the long-standing debate on the function of wall sculpturing in plant capillaries [14]. It is well known that wall sculpturing is common in plant species native to harsh environments, and enhanced wettability is likely to be of benefit in numerous regards to plants growing in such environments [16]. If wall sculpturing is indeed an adaptation designed to enhance wettability, further studies of such systems may provide biomimetic clues for the engineering of wettability in technological applications such as microfluidics, especially as the use of plastics, most of which are not perfectly wetted by water, becomes more common [21].

Measured values of both the relative frequency of occurrence of zero contact angles and the intrinsic contact angle θ_E for all six types of samples studied are reported in Fig. 4. In addition to the enhancement of wettability which arises from the presence of a pronounced warty layer in *C. endlicheri*, two further trends can be identified from these data. First, drying, or aging, leads to a reduction in wettability (reduced value of the frequency of zero contact angles, and a slight, but statistically significant, increase in equilibrium contact angle). This effect can be

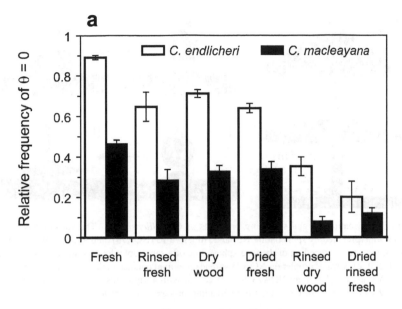

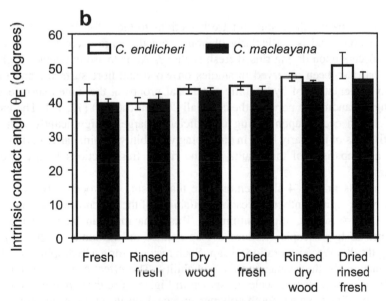

Figure 4. Wettability of the six different types of sapwood sections studied. Unfilled bars denote results for *C. endlicheri* (rough-walled capillaries); black bars denote results for *C. macleayana* (smooth-walled capillaries). (a) Relative frequency of zero contact angles (average ± standard error). (b) Intrinsic contact angle (average ± standard error).

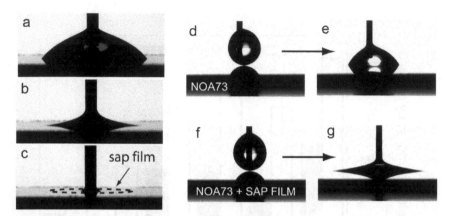

Figure 5. Formation of, and spreading on, films of *C. endlicheri* sap. The substrate is a smooth polymer film (prepared by spin coating Norland Adhesive 73 (NOA73) onto a silicon wafer, followed by uv-curing). (a–c) Receding of the three-phase line of a sap droplet leads to the formation of a sap film. (d) and (e) A water droplet on NOA73 spreads to a final contact angle of approximately 64°. (f) and (g) A water droplet on an NOA73 surface coated with a sap film spreads to final contact angle close to zero ($< 10°$). In both cases the spreading process is complete within approximately one second.

seen by comparing the results for fresh sections to those for dry-wood and dried-fresh sections, and is particularly obvious from the large decrease in wettability which occurs upon drying rinsed-fresh sections. An increase in contact angle with aging has also been observed in studies on wood and fibre surfaces, and may be due to reorientation of hydrophilic functional groups at the lumen surface and/or closing of nanometric pores in the cell wall ("hornification") [12, 13]. The increase in θ_E which occurs upon drying is sufficiently large to significantly reduce the effectiveness of the warty layer in enhancing wettability (again, particularly obvious from a comparison of the results for the rinsed fresh sections before and after drying).

The results in Fig. 4 also demonstrate that rinsing sapwood sections in excess (pure) water significantly reduces the wettability of the lumen walls, an observation which at first seems counterintuitive. This behaviour can be attributed to the presence of hygroscopic residues of sap solutes on the lumen walls, deposited during the drainage/evaporation of sap. Such residues form a hydrophilic coating of the lumen walls, thus enhancing the wettability. This effect can be illustrated very simply on a macroscopic scale, as shown in Fig. 5. The three-phase line of a sap droplet receding from a smooth polymer surface deposits a film of sap (for receding rates above $\sim 10^{-2}$ mm s^{-1}) (Fig. 5a–c). A water droplet which is subsequently contacted with the dried sap film spreads to a very low contact angle (Fig. 5f, g); in contrast, a water droplet contacted with the uncoated polymer surface spreads to a contact angle of approximately 64° (Fig. 5d, e). It should also be noted that

water films   water droplets

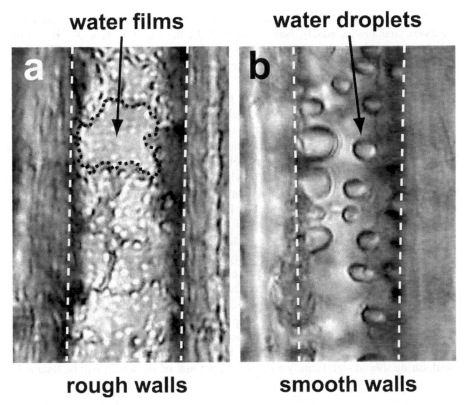

rough walls   smooth walls

Figure 6. Condensation of water on lumen walls during rehydration of (fresh) sapwood tracheids. (a) In tracheids of *C. endlicheri* water condenses as films, spanning clusters of warts. (b) In tracheids of *C. macleayana* water condenses as droplets.

the contact angles of pure water and sap on the *uncoated* polymer film were very similar (results not shown).

The presence of hygroscopic sap residues in dehydrated tracheids was obvious from microscopic observations of the rehydration of sapwood sections. During re-wetting of unrinsed sections, these residues absorb water, leading to condensation on lumen walls contained within air bubbles, as illustrated in Fig. 6. In tracheids of *C. endlicheri*, water condenses as films which span clusters of warts (Fig. 6a). In *C. macleayana*, water condenses as droplets (Fig. 6b), which grow with time and eventually bridge the tracheid diameter, leading to a 'pinching-off' of the original bubble into numerous smaller bubbles. Sap residues are removed when sapwood sections are rinsed in excess water, and wall condensation is no longer observed. It is known that sap contains many types of solutes [22], but regardless of the precise composition of the residues observed in these experiments, it is clear that they are very hygroscopic, as revealed by experiments in which sections were flooded with saturated salt solutions. Wall condensation was observed with water, and saturated aqueous solutions of $BaCl_2$, NaBr, and $NaNO_2$, but not with K_2CO_3. Based on tabulated values of the equilibrium humidity of saturated salt solutions [23], these observations suggest that the deliquescence point (humidity) of the residues is approximately 50%.

The presence of such residues clearly enhances wettability (Fig. 4), but is also likely to have a significant effect on the absorption of water by wood at humidities higher than 50%. Residual solutes may also aid in the refilling of embolised conduits in living plants, by ensuring the presence of water into which osmotica could be actively pumped from living cells [4–6], and by contributing to the total amount of dissolved solutes. It has previously been hypothesised that sap solutes (mucilage) may be beneficial in maintaining water transport in salt-tolerant plants [11, 24], but such effects may well be of significance in a broader range of plant species.

The amount of sap residue present on lumen walls will obviously depend on the manner in which the xylem conduit is dehydrated. A large fraction of the conduits in wood samples dried in air must contain relatively large amounts of solutes, because all of the solutes contained in the original sap must remain in the wood. In the cavitation of an individual conduit in a living plant, the amount of solutes remaining will be determined by the amount of sap left behind during the drainage process. The amount of residual sap will increase with the rate of drainage of the conduit [25], and with the surface roughness of the lumen walls [26]. In particular, wall sculpturing should significantly increase the amount of sap which will be retained.

4. CONCLUSIONS

The lumen walls of tree capillaries are not intrinsically perfectly wettable by water. The intrinsic, or equilibrium contact angle of water on (fresh) smooth lumen walls is approximately 40°, with a contact angle hysteresis of $\theta_A - \theta_R \approx 60°$. Prominent wall sculpturing leads to perfect wetting, which may explain why wall sculpturing is common in plant species native to harsh environments, where enhanced wettability is likely to be of benefit. Air drying increases the intrinsic contact angle by up to 10°, and significantly diminishes the effect of wall sculpturing on wettability. Finally, wettability is enhanced by the presence of sap solutes deposited on lumen walls during drainage of sap from conduits. These residues are hygroscopic and absorb significant amounts of water at humidities as low as 50%, an effect which may be of importance in understanding embolism repair, as well as the absorption of water by wood.

Acknowledgements

The author thanks V. S. J. Craig, R. D. Heady and P. D. Evans for discussions and encouragement, and N. Krause for providing the NOA 73 films. Thanks are also extended to the Australian National Botanic Gardens for granting me a permit (2004/06947) to collect Callitris samples, and the Australian Research Council for funding.

REFERENCES

1. M. T. Tyree and M. H. Zimmermann, *Xylem Structure and the Ascent of Sap*, 2nd edn. Springer, Berlin (2002).
2. W. F. Pickard, *Prog. Biophys. Mol. Biol.* **37**, 181 (1981).
3. A. M. Lewis, V. D. Harnden and M. T. Tyree, *Plant Physiol.* **106**, 1639 (1994).
4. N. M. Holbrook and M. A. Zwieniecki, *Plant Physiol.* **120**, 7 (1999).
5. M. A. Zwieniecki and N. M. Holbrook, *Plant Physiol.* **123**, 1015 (2000).
6. U. G. Hacke and J. S. Sperry, *Plant Cell Environ.* **26**, 303 (2003).
7. W. Konrad and A. Roth-Nebelsick, *J. Biol. Phys.* **31**, 57 (2005).
8. S. Malkov, V. Kuzmin, V. Baltakhinov and P. Tikka, *J. Pulp Paper Sci.* **29**, 137 (2003).
9. M. de Meijer, K. Thurich and H. Militz, *Holz Roh Werkst.* **59**, 35 (2001).
10. R. Laschimke, *Thermochim. Acta* **151**, 33 (1989).
11. U. Zimmermann, H. Schneider, L. H. Wegner and A. Haase, *New Phytol.* **162**, 575 (2004).
12. G. I. Mantanis and R. A. Young, *Wood Sci. Technol.* **31**, 339 (1997).
13. M. Wålinder, Wetting phenomena on wood, PhD Thesis, Division of Wood Technology and Processing, Royal Institute of Technology, Stockholm (2000).
14. S. J. Carlquist, *Comparative Wood Anatomy: Systematic, Ecological, and Evolutionary Aspects of Dicotyledon Wood*. Springer, Berlin (1988).
15. R. D. Heady, R. B. Cunningham, C. F. Donnelly and P. D. Evans, *IAWA J.* **15**, 265 (1994).
16. M. M. Kohonen, *Langmuir* **22**, 3148 (2006).
17. C. Andrieu, C. Sykes and F. Brochard, *Langmuir* **10**, 2077 (1994).
18. D. Quéré, *Rep. Prog. Phys.* **68**, 2495 (2005).
19. G. E. Fogg, *Proc. R. Soc. B* **134**, 503 (1947).
20. W. Barthlott and C. Neinhuis, *Planta* **202**, 1 (1997).
21. H. Becker and L. E. Locascio, *Talanta* **56**, 267 (2002).
22. B. Sattelmacher, *New Phytol.* **149**, 167 (2001).
23. R. C. Weast (Ed.), *CRC Handbook of Chemistry and Physics*, 67th edn. CRC Press, Boca Raton, FL (1986).
24. P. J. Melcher, G. Goldstein, F. C. Meinzer, D. E. Yount, T. J. Jones, N. M. Holbrook and C. X. Huang, *Oecologia* **126**, 182 (2001).
25. D. Quéré and E. Archer, *Europhys. Lett.* **24**, 761 (1993).
26. E. T. Choong and F. O. Tesoro, *Wood Sci. Technol.* **23**, 139 (1989).

Contact Angle, Wettability and Adhesion, Vol. 5, pp. 59–72
Ed. K.L. Mittal
© VSP 2008

The effect of binary fatty acids systems on partitioning, IFT and wettability of calcite surfaces

O. KAROUSSI and A. A. HAMOUDA*

Department of Petroleum Technology, University of Stavanger, P.O. Box 8002 Ullandhaug, NO-4036 Stavanger, Norway

Abstract—The effect of binary systems composed of 18-phenyloctadecanoic acid (PODA) and each of the following acids: heptanoic (HPA), stearic (SA), and oleic (OA) on the wettability alteration of calcite, IFT and partitioning coefficient is investigated. For SA and OA in single systems in n-C10/water, the interfacial pKa is estimated to be at about pH7. While in single system of PODA in n-C10/water, the interfacial pKa is shown to be at a pH > 8. For binary systems of PODA/SA and PODA/OA, the interfacial pKa values for both systems are at a pH > 8, indicating the influence of PODA on the binary system. This may be explained based on hydrophobic intermolecular interaction. However, for PODA/HPA binary system, PODA showed no effect on the pKa. Since HPA is more soluble in the water phase, at high pH a possible repulsion of the ionized molecules occurs. This perhaps reduces/inhibits a complete dissociation of PODA, hence interfacial pKa was not observed.

The presence of PODA caused SA or OA affinity more toward the oil phase whereas for HPA (0.01 M) it was more toward the water phase. The increase of the partitioning of HPA (0.002 M) in the binary system toward the oil phase may be explained based on the proposed hydrophobic interaction between the acids in the binary system.

The measured contact angles in the binary systems are close to the measured contact angles for PODA in a single system. This indicates the dominance of PODA on the calcite surface.

Like a single fatty acid, the binary system shows its dependence on the pH in terms of IFT, K (partitioning coefficient) and contact angle. It can be seen that there is a decrease of these parameters as the pH increases.

Keywords: Wettability (contact angle); partitioning coefficient; interfacial tension (IFT); calcite; 18-phenyloctadecanoic acid (PODA); stearic acid (SA); oleic acid (OA); heptanoic acid (HPA).

1. INTRODUCTION

Polar components in crude oils have been recognized for their contribution to wettability alteration of reservoir rocks. Organic fatty acids (RCOOH) in presence of a water film are found to modify carbonate rocks [1–7]. The interfacial tension of oil/water system in an oil reservoir as well as distribution of components in oil or water phase (partitioning), addressed by Hamouda and coworkers [1], play

---

*To whom correspondence should be addressed. Tel.: +47 51 83 22 71; Fax: +47 51 83 17 50; e-mail: aly.hamouda@uis.no

important roles in the wetting state of the calcite surfaces. In terms of partitioning coefficient of fatty acids, a decrease with pH is observed. Standal *et al.* [8] showed a relationship between the distribution of the polar component and IFT.

Most of the investigations of model components in an oil/water system in the literature are related to the effect of single component of polar functional group. Hamouda and coworkers [1, 2], Spildo and coworkers [7, 9], Standal *et al.* [8], Zullig and Morse [10], Madsen *et al.* [11], and Vodnár *et al.* [12] have studied single system additives. However, for binary systems, Rudin and Wasan [13] mentioned only briefly in their work about the interfacial activity of oleic and lauric acids. Spildo and coworkers [14, 15] worked with homologous binary systems of octanoic acid and octylamine representing organic acid and base. Zhang *et al.* [16] and Zhang *et al.* [17] have investigated the fatty acid/surfactant and surfactant/surfactant binary systems, respectively.

In this work the effects of binary systems on partitioning in n-C10/water system, IFT and wettability of calcite are investigated. Binary systems, addressed here, mainly composed of stearic acid (SA), oleic acid (OA) or heptanoic acid (HPA) with 18-phenyloctadecanoic acid (PODA) in n-C10.

2. EXPERIMENTAL

2.1. Materials

The effect of the purity of n-decane is addressed here for a comparison between 99 and 95% n-decane. Both n-decane grades are produced by Merck and supplied by VWR International AS, Norway. The work presented here for IFT was carried out using 95% grade n-decane. Chiron AS, Norway supplied PODA, and SA (>98.5%), OA (>98%) and HPA (>99%) were obtained from Fluka, AS, Norway. In this work, n-decane represents the oil phase and hereafter will be termed "oil phase" unless otherwise stated. The aqueous phase used was distilled water.

The calcite crystals, used in contact angle measurements, were "Island-spar" calcite from India, supplied by J. Brommeland AS, Norway. The preparation method for calcite samples was the same procedure as reported previously [1].

2.2. Methods

2.2.1. pH measurements
The pH measurements were performed by a pH meter model PHM 92 (supplied by Radiometer Analytical A/S, Denmark). The aqueous solutions were adjusted to the selected values using drops of 0.05 M NaOH and 0.05 M HCl. The pH values in this study were between 4 and 10. The accuracy of pH measurement was about ±0.05.

2.2.2. Partitioning coefficient (K) determination
2.2.2.1. Single acid systems. Partitioning coefficients of low (0.002 M) and high (0.01 M) concentrations of single fatty acids dissolved in n-decane were determined

following the method described earlier [1]. The only difference was that a rotator (25 rpm) was employed instead of a shaker in the previous work for agitating the oil and water. The effect of rotating instead of shaking on oil/water mixing was checked by measuring the IR absorption value of equilibrated oil phase for the two systems, and no significant difference was observed. Standard deviation of the measured (log K) value was estimated to be about 0.15 (based on 5 measurements).

2.2.2.2. Binary acid systems. In general, a binary system consisted of 0.01 M PODA in n-decane with concentrations of 0.002 or 0.01 M of SA, OA or HPA, referred to hereafter as "secondary acids". First, a calibration chart (IR absorption vs concentration, 0.002, 0.005, 0.008 and 0.01 M in n-C10) for PODA was established. Then, calibration charts for binary systems were established after addition of the secondary acids with concentrations of 0.002, 0.005, 0.008 and 0.01 M dissolved in 0.01 M PODA/n-decane system. Assuming that PODA has more propensity than other acids to remain in the oil phase (i.e. higher partitioning coefficient), a corresponding absorption value for single 0.01 M PODA is then deducted from binary calibration chart to construct a new calibration chart for each secondary acid. In other words, the new calibration chart reflects the effect of PODA on the partitioning of the secondary acid. Using the new calibration chart and equations (1) and (2), the partitioning coefficient "K" for PODA binary system is obtained. The results are discussed in Section 3.2.

$$C_{w,eq} = C_{o,i} - C_{o,eq} \tag{1}$$

$$K = \frac{C_{o,eq}}{C_{w,eq}} \tag{2}$$

where, $C_{o,i}$ and $C_{o,eq}$ are the initial and equilibrium concentrations of fatty acid in the oil phase, respectively. $C_{w,eq}$ is the concentration of the acid in the water phase at equilibrium.

2.2.3. Interfacial tension measurements
The interfacial tension (IFT) measurements were performed by the ring method, using a KRÜSS tensiometer type 8451, with an accuracy of ±1.5 mN/m.

2.2.4. Contact angle measurements
The same procedure as reported earlier [1] was followed for the preparation of the pieces of calcite for the contact angle measurements. Contact angle measurements were carried out at pH of 6.6, 8 and 10. The measured contact angles were accurate to ±3°.

3. RESULTS AND DISCUSSION

3.1. Interfacial tension measurements for binary systems

The effect of n-decane purity on IFT was investigated. This was done for 0.01 M PODA/n-decane (95 and 99%)/water systems. As shown in Fig. 1, no difference was found in the measured IFT between 95 and 99% n-decane. A similar test was also done for other acids with the same conclusion. It was, therefore, decided to use 95% n-decane for both economical and practical reasons.

For single systems, the highest reduction in IFT was obtained by PODA compared to the other fatty acids tested (Fig. 2). This suggests that the order of the interfacial activity for these acids is as follows: PODA > OA > SA >HPA.

Lowering of IFT due to the presence of a surface-active component like carboxylic acid in an acidic oil/alkali system has been widely studied by many researchers. Rudin and Wasan [13] suggested that the decrease in IFT was related to the adsorption of ionized acid molecules onto the interface along with un-ionized acid molecules. They stated that regardless of acid hydrocarbon chain length, IFT goes thorough a minimum with pH. Hoeiland et al. [5], Standal et al. [8], Spildo and Høiland [9] and Touhami et al. [18] also confirmed the decrease in IFT due to ionization of acid molecules at high pH. In this work, it is observed that PODA in n-decane/water system gives the lowest IFT at pH 4 (24 ± 1.5 mN/m) in single system. For both SA and OA in n-decane/water systems at pH 4 the measured IFT values are 31.9 and 28.5 ± 1.5 mN/m, respectively. For pH $\geqslant$ 8, only small difference in IFT measurements for OA and PODA was observed. However, at pH 10, IFT for SA (in

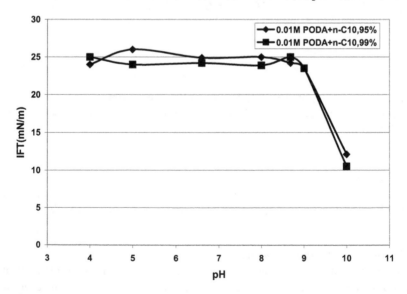

Figure 1. A comparison between PODA IFT in n-C10/water system for 95 and 99% purity of n-C10.

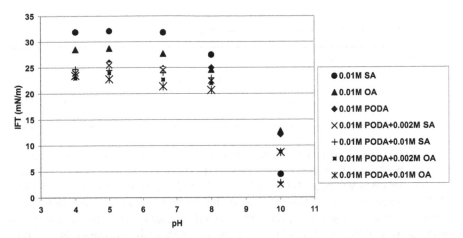

Figure 2. Comparison between IFT of dissolved single (PODA, SA and OA in n-C10) and binary fatty acids (PODA/SA or OA) in n-decane/water systems as a function of pH.

single system) declined to 4.5 ± 1.5 mN/m. It is worth mentioning that at pH 10, a layer at the interface is formed. This, most likely, is a salt formed by the reaction of SA with Na^+, which is added in the form of NaOH for the pH adjustment. It is estimated that the concentration of sodium ions in NaOH used to adjust the pH to 10 is about 0.001 M. The literature [19] shows that sodium stearate is soluble in water, which may confirm our interpretation that salt formation contributed to the further reduction of IFT, which would otherwise be only due to the complete SA dissociation. If one attempts to quantify the contribution of the salt, assuming that IFT for completely dissociated acid in this system is about 12.8 ± 1.5 mN/m (which is equivalent to OA IFT @pH 10), about three times reduction in IFT is obtained when salt (sodium stearate) is formed.

A comparison between IFT as a function of pH for binary systems (0.01M PODA in n-C10 with 0.01 M of SA or OA/water) is shown in Fig. 2. Two main observations can be made from this figure. The first observation is the synergism in the binary systems studied, where a further reduction of IFT from the corresponding single systems below pH 8 (e.g. at pH 4, 31.9 and 28.5 ± 1.5 mN/m, for 0.01 M SA and OA, respectively) to about 24 mN/m takes place, which almost corresponds to the IFT of PODA single system (PODA in n-C10/water system). The second observation is that the order of the IFT reduction in the binary systems follows the same order as that for the single systems, i.e. SA > OA for pH between 4 and 8. However, above pH 8, for SA at both low (0.002 M) and high (0.01 M) concentrations the same IFT is obtained. There is no difference within the experimental error of ± 1.5 mN/m. OA shows a consistently distinct, but not large, IFT difference between the low and high concentrations, 0.002 and 0.01 M, where more reduction of IFT is observed with the higher concentration for pH between 4 and 8.

Table 1.
Surface molecular areas for fatty acids in single and binary systems at pH 6.6

Molecular area (Å²/molecule)						
Single				Binaries (Equimolar concentrations)		
PODA	SA	OA	HPA	PODA + SA	PODA + OA	PODA+HPA
48	58	66	88	37	45	46.6

It may be concluded that the binary systems (PODA with SA or OA) tested so far have been strongly influenced by PODA. In other words, the IFT of the binary systems (secondary fatty acids (i.e. SA and OA) added to 0.01 M PODA), regardless of their concentrations, follow the same IFT trend as observed for PODA. It is interesting that during the experiments, no salt formation (sodium stearate) was observed in PODA/SA system. In addition, although a slight decrease is observed in the IFT with increasing pH, the interfacial pKa seems to occur for SA and OA at a pH > 8.0 similar to PODA. An interfacial pKa of 8.8 is estimated for both single 0.01 M PODA-n-decane/ water system and its binaries with 0.01 M SA and 0.01 M OA. The above domination of PODA in interfacial activity of binary system was further investigated by estimating the occupied area by the different fatty acid molecules, using the Gibbs adsorption equation:

$$\Gamma_i = -\frac{1}{RT}\left(\frac{\partial \gamma}{\partial \ln C_i}\right) \tag{3}$$

where Γ_i is surface excess concentration, R is universal gas constant (i.e. 8.314 J/Mole.K); T is temperature (K), C is total bulk concentration and γ is surface (interfacial) tension (mN/m).

The calculated surface molecular areas for single and binary fatty acid systems at pH = 6.6 (i.e. distilled water with no added ion for adjusting the pH) are given in Table 1. As indicated by the surface areas, PODA possesses the closest packed molecules at the interface followed by SA, OA and HPA. Equimolar concentrations of PODA and added secondary acids (0.01/0.01 M) at a pH of 6.6 were used in the calculations.

A binary system with a relatively short chain fatty acid with PODA was also investigated. In this system 0.002 and 0.01 M of HPA was tested in the presence of 0.01 M PODA in n-C10/water system. This is shown in Fig. 3. In this figure IFT of the single system of HPA (0.01 M HPA in n-C10) is compared with the binary system. As can be seen, the pH has no effect on IFT, as shown in our earlier work for a single system [1]. Similar to other fatty acids, SA and OA, IFT for the binary system of PODA/HPA is reduced from ≅34 mN/m (for HPA single system) to a value in the vicinity of the IFT for PODA (for a single system), which is about 24 mN/m at a pH below 8. However, it is interesting to observe that the binary system (PODA/HPA) does not show the anticipated further reduction in IFT at the interfacial pKa for PODA in the single system at pH > 8. Zhang

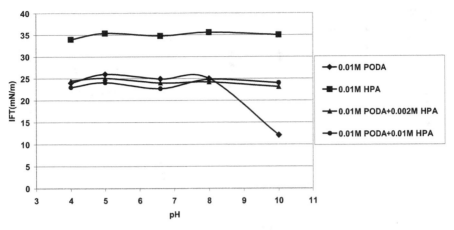

Figure 3. IFT of dissolved PODA, HPA and their binaries in n-decane/water system as a function of pH.

et al. [16] investigated the effect of a binary system of 5-phenylalkane sulfonate (as an active surfactant, which is reported to reduce IFT to about 0.1 mN/m) and oleic acid on IFT. Their conclusion was that when oleic acid was added to distilled water containing NaOH, sodium oleate was formed by the reaction between the oleic acid and NaOH but they did not mention exactly at what pH the oleate formation occurred. They showed that the presence of oleate at the interface inhibited the adsorption of 5-phenylalkane sulfonate, hence a lower concentration of 5-phenylalkane sulfonate at the interface, and consequently a higher IFT in the presence of oleate was measured.

In order to further address this phenomenon observed in the binary system of PODA/HPA, measurement of IFT in SA/OA in n-C10/water binary system was performed as a function of pH. The results are shown in Fig. 4a, illustrating that even at low concentration of OA (0.002 M) added to 0.01 M SA, the IFT is reduced following the same trend as that for the single system of 0.01 M OA. Higher added concentration of OA (0.01 M) to 0.01 M SA caused further reduction of IFT as expected. It can be seen that with a low concentration of OA (0.002 M), the IFT at pH > 8 has almost the same slope as that for SA (single system), while at higher concentration of OA (0.01 M), slightly less steep reduction in IFT at pH > 8 is shown.

The effect of HPA on SA was also studied for the binary system of SA/HPA. As illustrated in Fig. 4b, a similar behavior to the binary system of PODA/HPA is observed, where the interfacial activity of the binary system is increased. Again in the presence of HPA, increasing the pH did not show any further reduction in IFT, following the same trend as that for HPA but at a lower IFT ($\cong$ 29.5 mN/m) than that for HPA single system ($\cong$ 35 mN/m).

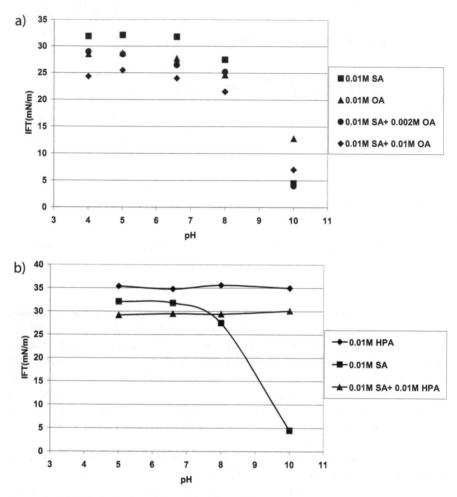

Figure 4. (a) IFT of dissolved SA or OA and their binary systems in n-decane/water system as a function of pH. (b) IFT of binary system (HPA+SA/n-C10)/water as a function of pH.

3.2. Partitioning coefficient

Figure 5 shows that PODA has the highest partition coefficient among all the acids used in this study. Hamouda and coworkers in previous works [1, 20] concluded that a fatty acid with a higher partitioning coefficient (K) shows a lower interfacial tension. They ranked the fatty acids studied based on partitioning coefficient in n-C10/water system as follows: PODA>SA>OA>HPA. They also demonstrated the dependence of K on pH where, as expected, K is reduced with increase in pH. Spildo and Høiland [9] have also reported a similar trend for single fatty acid systems.

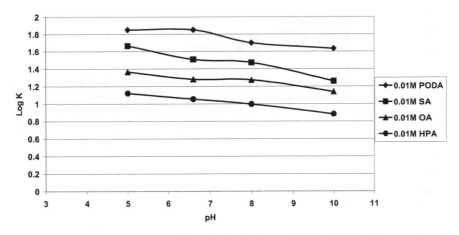

Figure 5. Logarithmic partition coefficient for single fatty acids dissolved in n-decane and equilibrated with distilled water at different pH values.

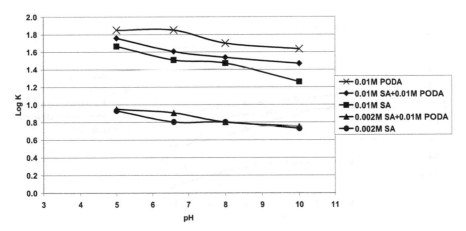

Figure 6. Logarithmic partition coefficient for SA in absence (single system) and presence of PODA (binary system) dissolved in n-decane and equilibrated with distilled water at different pH values.

Partitioning coefficients (presented here as log K, for convenience and simplicity, but will be termed as K hereafter, unless otherwise stated) of the secondary added fatty acids in binary systems of PODA/SA, PODA/OA and PODA/HPA for two concentrations (0.002 M and 0.01 M) of the secondary fatty acids (SA, OA and HPA) are shown in Figs. 6, 7 and 8, respectively while the concentration of PODA was kept constant in all experiments (0.01 M). As can be seen, for equimolar concentration of SA and PODA (0.01 M), the K of SA in the binary system is between the K for PODA and that for SA (Fig. 6), and is rather closer to the K

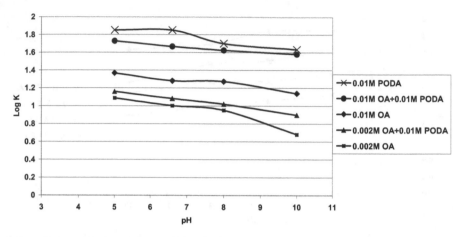

Figure 7. Logarithmic partition coefficient for OA in absence (single system) and presence of PODA (binary system) dissolved in n-decane and equilibrated with distilled water at different pH values.

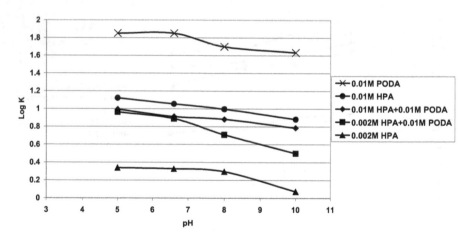

Figure 8. Logarithmic partition coefficient for HPA in absence (single system) and presence of PODA (binary system) dissolved in n-decane and equilibrated with distilled water at different pH values.

for SA (single system) than that for PODA (single system), except at pH of 10. It is interesting to see that at low SA concentration (0.002 M), the partitioning coefficient shifted more towards the water phase for both single and binary systems and the K of SA for the binary system is almost the same as the K of SA in single system. Similar trend is observed for OA (Fig. 7), where the K of the equimolar concentration (0.01 M) of OA in the binary system is between the K for PODA single system and that for OA single system; however, unlike SA it is rather closer to the K for PODA at all pHs. At 0.002 M of OA in the binary system, again the

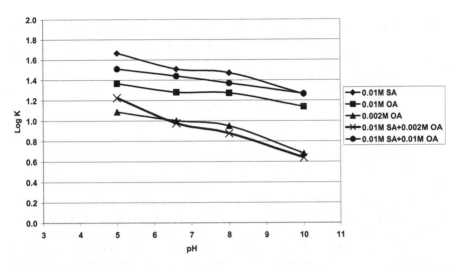

Figure 9. Logarithmic partition coefficient for OA in absence (single system) and presence of SA (binary system) dissolved in n-decane and equilibrated with distilled water at different pH values.

K for OA is decreased (hence increased concentration in the water phase), but to a lesser extent than that for SA. Figure 8 shows that in case of HPA, the K for HPA in the binary system is between the K for HPA single system at both 0.01 and 0.002 M, but closer to the K for HPA in the binary system at 0.01 M.

In summary, it seems that in the binary systems PODA/SA and PODA/OA, 0.01 M of PODA influences the partitioning of both SA and OA more towards the oil phase for the high concentration (0.01 M of SA and OA), while there is less effect of PODA on SA and OA at the low concentration (0.002 M of SA and OA) as shown in Figs. 6 and 7. This may be explained in terms of the hydrophobic intra- and inter-molecular bonding between the long chain molecules i.e. PODA/SA and PODA/OA, hence reducing the partitioning of these two acids. This is evidenced from the change in the molecular packing at the interfacial boundary. For example, the occupied molecular areas for SA and OA in the binary system with 0.01 M PODA have been compressed from 58 to 37 and from 66 to 45 $Å^2$/molecule, respectively. At the low concentration, on the other hand, the interfacial area occupied by PODA could be accommodated with low concentration of the added secondary acids without a large change in the surface pressure, hence negligible effect at low concentration. Zullig and Morse [10] proposed an explanation for co-adsorption of myristate and stearate on a mineral surface that led to a substantial increase in myristate adsorption as a result of hydrophobic bonding between myristate and adsorbed stearate.

In the case of the relatively short chain fatty acid (HPA), as shown earlier in Fig. 3, in the presence of PODA IFT was reduced, which may indicate a reduction in the interfacial energy. This becomes evident from the molecular packing of HPA, where the occupied area by a molecule is reduced from 88 to about 46 Å^2/molecule, i.e. by almost 50%. Although, this seems at first glance to contradict the results obtained by the partitioning tests, it does confirm the reduction of the energy barrier enhancing the affinity of the HPA towards the water, hence decreasing the partitioning coefficient of HPA. In the case of the low concentration HPA (0.002 M) single system, it is perhaps close to the solubility of HPA in the water as indicated in Fig. 8. The increase of the partitioning of HPA (0.002 M) in the binary system toward the oil phase may be explained based on the proposed hypothesis, being due to the hydrophobic interaction between the acids in the binary system. As the pH increases above $\cong 7$ more HPA dissociation occurs, hence higher reduction of the K values.

The degree of reduction of K at pH > 7 is lower in the case of higher concentration HPA than that with low concentration, which supports the above hypothesis and further demonstrates the effective balance between hydrophobic intermolecular interaction and repulsion between the alike charges. The latter becomes more evident at lower concentration of HPA, where ionization and higher solubility of HPA in water may overcome the intermolecular hydrophobic interaction between PODA and HPA, because of short chain of HPA.

This hypothesis was further tested with a binary system of SA/OA at a constant concentration of SA (0.01 M) and two concentrations of OA (0.002 M and 0.01 M). As shown in Fig. 9, the partitioning coefficient of OA has increased for both concentrations at pH 5 and then it continues to decrease as the pH increases. Similar to the influence of PODA on SA or OA, the effect of 0.01 M SA on 0.01 M OA has increased the affinity of 0.01 M OA toward the oil phase, while 0.01 M SA has shown insignificant effect on low concentration of OA (0.002 M).

3.3. Contact angle measurement

The measured contact angles for PODA (in single system), and its binary systems with SA and HPA are shown in Fig. 10. The higher measured contact angles for SA than PODA indicate that SA modifies the calcite surface to a more oil-wet than PODA. In this work the advancing contact angles were about 153° and 125° for 0.01 M SA and 0.01 M PODA at pH 6.6, respectively. Hamouda and Rezaei Gomari reported [20] an advancing contact angle of about 160° for SA at pH 6.5.

The measured contact angles in binary systems, especially for PODA-HPA, are close to the measured contact angles for PODA in single system. This indicates the dominance of PODA on the calcite surface. The measured contact angle of PODA-OA binary system at pH = 8 is shown to be $115 \pm 3°$. This is close to that of PODA in single system at pH 8.

In order to re-check the dominance of PODA in the binary system, some pre-wetted calcite chunks with distilled water at pH = 8 were modified with 0.01 M SA for 24 hours and dried in a closed system for 24 hours. Then the modified sample again was humidified at the same pH as done for SA and immersed into 0.01 M PODA solution. After the drying process, advancing contact angle of $127 \pm 3°$ was

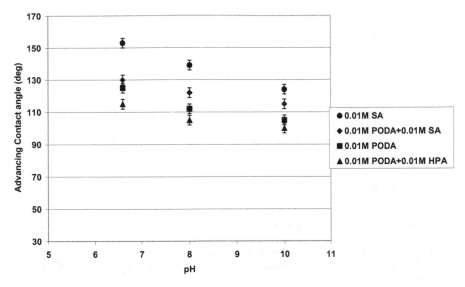

Figure 10. Measured advancing contact angles for single SA (0.01 M), single PODA (0.01 M) and its binary systems with SA (0.01 M) and HPA (0.01 M) on calcite.

measured which is close to the contact angle on treated calcite surfaces with the SA-PODA binary (i.e. 122 ± 3°) at the same pH and concentration.

4. CONCLUSIONS

PODA influences the other acids co-present in binary systems in terms of both interfacial tension and partitioning coefficient.

For binary systems of PODA/SA and PODA/OA, the interfacial pKa for both are at a pH > 8 indicating the influence of PODA on the binary system. This may be explained based on hydrophobic intermolecular interaction. However, for PODA/HPA binary system, PODA showed no effect on the pKa. Since HPA is more soluble in the water phase, at high pH a possible repulsion of the ionized molecules occurs. This perhaps reduces/inhibits a complete dissociation of PODA, hence interfacial pKa was not observed.

The presence of PODA caused higher affinity of SA or OA toward the oil phase whereas for HPA (0.01 M) it was higher toward the water phase. The increase in the partitioning of HPA (0.002 M) in the binary system toward the oil phase may be explained based on the proposed hypothesis, i.e., due to the hydrophobic interaction between the acids in the binary system.

The measured contact angles in binary systems are close to the measured contact angles for PODA in single system. This indicates the dominance of PODA on the calcite surface.

Like a single fatty acid, the binary systems show their dependence on pH in terms of IFT, K and contact angle.

Acknowledgements

The authors would like to acknowledge the University of Stavanger for the financial support of this project. We would also like to thank Karam Gomari, Ph.D. student in our group for his useful advice in the experimental part of this work and Unni Haukli for her positive attitude and help in getting the chemicals needed in time.

REFERENCES

1. K. A. Rezaei Gomari, A. A. Hamouda, T. Davidian and D. A. Fargland, in: *Contact Angle, Wettability, and Adhesion*, Vol. 4, K. L. Mittal (Ed.), pp. 351-367. VSP/BRILL, Leiden, The Netherlands (2006).
2. K. A. Rezaei Gomari, R. Denoyel and A. A. Hamouda, *J. Petrol. Sci. Eng.* **50**, 140-150 (2006).
3. K. A. Rezaei Gomari, R. Denoyel and A. A. Hamouda, *J. Colloid Interface Sci.* **297**, 470-479 (2006).
4. G. Hansen, A. A. Hamouda and R. Denoyel, *Colloids Surfaces* **172**, 7-16 (2000).
5. S. Hoeiland, T. Barth, A. M. Blokhus and A. Skauge, *J. Petrol. Sci. Eng.* **30**, 91-103 (2001).
6. K. Kowalewski, T. Holt and O. Torsaeter, *J. Petrol. Sci. Eng.* **33**, 19-28 (2002).
7. K. Spildo, H. Høiland and M. K. Olsen, *J. Colloid Interface Sci.* **221**, 124-132 (2000).
8. S. H. Standal, A. M. Blokhus, J. Haavik, A. Skauge and T. Barth, *J. Colloid Interface Sci.* **212**, 33-41 (1999).
9. K. Spildo and H. Høiland, *J. Colloid Interface Sci.* **209**, 99-108 (1999).
10. J. J. Zullig and J. W. Morse, *Geochim. Cosmochim. Acta* **52**, 1667-1678 (1988).
11. L. Madsen, L. G. Madsen, C. Grøn, I. Lind and J. Engell, *Org. Geochem.* **24**, 1151-1155 (1996).
12. J. D. Vodnár, M. Sălăjan and D. A. Lowy, *J. Colloid Interface Sci.* **183**, 424-430 (1996).
13. J. Rudin and D. T. Wasan, *Colloids Surfaces* **68**, 67-79 (1992).
14. K. Spildo, A. M. Blokhus and A. Andersson, *J. Colloid Interface Sci.* **243**, 483-490 (2001).
15. K. Spildo, A. M. Blokhus and A. Andersson, *J. Colloid Interface Sci.* **252**, 470-472 (2002).
16. S. Zhang, J. Yan, H. Qi, J. Luan, W. Qiao and Z. Li, *J. Petrol. Sci. Eng.* **47**, 117-122 (2005).
17. L. Zhang, L. Luo, S. Zhao and J. Yu, *J. Colloid Interface Sci.* **251**, 166-171 (2002).
18. Y. Touhami, V. Hornof and G. H. Heale, *J. Colloid Interface Sci.* **177**, 446-455 (1996).
19. D. R. Lide (Ed.), *Handbook of Chemistry and Physics*, 73rd ed, pp. 4-101. CRC Press, Boca Raton, FL (1992-1993).
20. A. A. Hamouda and K. A. Rezaei Gomari, paper No.99848 presented at the SPE/DOE Symposium on Improved Oil Recovery held in Tulsa, Oklahoma, USA (2006).

Contact Angle, Wettability and Adhesion, Vol. 5, pp. 73–94
Ed. K.L. Mittal
© VSP 2008

Mechanistic modeling of dynamic vapor-liquid interfacial tension in complex petroleum fluids

SUBHASH C. AYIRALA* and DANDINA N. RAO†

The Craft & Hawkins Department of Petroleum Engineering, Louisiana State University, Baton Rouge, LA 70803-6417, USA

Abstract—Determination of vapor-liquid interfacial tension for crude oils is crucial in improved oil recovery processes such as gas and chemical injection, and in multiphase flow dynamics in porous media. In this paper, we have proposed a new mechanistic parachor model to predict dynamic vapor-liquid interfacial tension of crude oils by incorporating mass transfer effects into the traditional parachor model, using the ratio of vapor-liquid diffusivities raised to an exponent. The sign and magnitude of the exponent in the proposed model have been used to characterize the type and extent of the governing mass transfer mechanism (vaporization or condensation) responsible for attaining vapor-liquid phase equilibria.

Three different crude oils with measured vapor-liquid interfacial tension data have been investigated using the proposed model. The proposed model has been able to predict vapor-liquid interfacial tensions accurately for all the three crude oils considered. Positive exponents obtained for the diffusivity ratios in the mechanistic parachor model indicate that vaporization of components from crude oil into the vapor phase is the governing mass transfer mechanism responsible for attaining vapor-liquid phase equilibria in all these three systems. A generalized regression model has also been developed for *a priori* prediction of the mechanistic parachor model exponent based on crude oil composition and this regression model has been validated using the compositional data of Prudhoe Bay crude oil.

Keywords: Interfacial tension; crude oils; parachor; diffusivity; mass transfer; vaporization; condensation; fluid phase equilibria; mechanistic parachor model; equation of state.

1. INTRODUCTION

1.1. Vapor-liquid interfacial tension

Interfacial tension is defined as the tension that exists at the interface of the two fluid phases due to the imbalance between the molecular attractive and repulsive forces. The concept of capillary pressure is often used in reservoir engineering calculations to investigate the distribution of fluids in crude oil reservoirs. The capillary pressure

---

*Presently with Shell International E & P Inc., 3737 Bellaire Blvd., Houston, TX 77025, USA.

†To whom correspondence should be addressed. Tel.: 225-578-6037; Fax: 225-578-6039; e-mail: dnrao@lsu.edu

is a strong function of the interfacial tension between the fluid phases. Interfacial tension also determines the fluid phase relative permeabilities, which describe the flow behavior of hydrocarbon fluids in porous reservoir rocks. The important role played by interfacial tension in petroleum engineering applications has been well explored in the literature [1].

The importance of the estimation of vapor-liquid interfacial tension for crude oils, and especially the gas condensates, has been well recognized in the oil industry. Gas condensates are a special type of hydrocarbon fluids occurring in crude oil reservoirs and they account for a significant portion of hydrocarbon reserves in the world. In gas condensate reservoirs, retrograde condensation of hydrocarbon liquids takes place as the reservoir pressure falls below the dew-point pressure (the minimum pressure at which the first drop of liquid appears when the gas-phase is allowed to expand at reservoir temperature) due to production. The volumes of such dropped out liquids in reservoirs are generally small, close to the critical saturation and hence are considered to be immobile and unrecoverable. However, it is possible to improve the liquid phase mobility and reduce the critical saturation in the liquid dropped out regions of gas condensate reservoirs by altering the relative permeabilities. Gas condensate relative permeabilities are generally a function of capillary number, which is defined as the ratio of viscous to capillary forces. Interfacial tension thus plays an important role, through capillary forces, in developing accurate relative permeability-capillary number models that are useful for optimization of production strategy in gas condensate fields.

Vapor-liquid interfacial tensions are also essential for proper understanding of several pore level interactions taking place in crude oil reservoirs. It is largely known that gravity drainage is the main mechanism responsible for oil recovery in naturally fractured reservoirs. Prior knowledge of vapor-liquid interfacial tension is, therefore, useful to predict the capillary pressure and hence the oil recoveries due to gravity drainage in such fractured reservoirs.

The interfacial tension between fluid phases is particularly targeted by the petroleum engineers in the area of improved oil recovery using chemicals and miscible gas solvents. In miscible gas injection processes, improved oil recovery relies greatly on the interactions between the displacing gas and displaced oil to create near-zero interfacial tension between the fluid phases. Miscible displacements using CO_2 gas are currently becoming more popular in the United States due to their ability to recover higher amounts of oil and the potential for CO_2 sequestration. However, prior determination of miscibility conditions is essential for process design and economic success of such miscible gas injection projects in the field. Miscibility means zero interfacial tension between the fluid phases [2–4] and hence gas-oil interfacial tensions can be used to accurately determine miscibility conditions based on this fundamental definition [5–7]. Thus low interfacial tension between the fluid phases is desired for reduced capillary pressure and residual oil saturations and is the ultimate objective of any chemical and miscible gas injection improved oil recovery process.

1.2. Traditional vapor-liquid IFT prediction models

From the discussion provided in the previous section, it is evident that determination of vapor-liquid interfacial tension is crucial for developing gas condensate relative permeability-capillary number models, to predict gravity drainage oil recoveries in fractured reservoirs, in improved oil recovery processes such as gas and chemical injection, and in multiphase flow dynamics in porous media. However, the experimental data on interfacial tension for complex fluid systems such as crude oils involving multicomponents are scarce. This is mainly due to the difficulty in measuring interfacial tension for such complex systems, especially at demanding conditions of pressure and temperature that exist in petroleum reservoirs. As a result, several IFT prediction models that were proposed in the literature for pure-component hydrocarbons have been utilized to predict the interfacial tension of complex petroleum fluids. The most important among these models are the parachor model [8–10], the corresponding states theory [11], thermodynamic correlations [12], and the gradient theory [13].

The parachor model [8–10] is probably the oldest among all the IFT prediction models and is still widely used in the petroleum industry to estimate the vapor-liquid interfacial tension of crude oils. Empirical density correlations are used in this model to predict the interfacial tension. The parachor model is expressed by,

$$\sigma^{1/4} = \rho_M^L \sum x_i P_i - \rho_M^V \sum y_i P_i \tag{1}$$

where σ is the interfacial tension in mN/m, ρ_M^L and ρ_M^V are the molar densities of the liquid and vapor phases, respectively, in gmole/cm^3 and x_i and y_i are the mole fractions of component i in the liquid and vapor phases, respectively, and P_i is the parachor of the component i. The parachor values of different pure compounds have been determined from measured surface tension data [14] and are used in parachor model based IFT calculations. A detailed discussion on the 'parachor' concept has been further provided in Section 3.4 to understand the underlying physics and thermodynamics behind the parachor.

The parachor model has been extensively tested for predicting surface tension of pure components and binary mixtures. The match between the parachor model predictions and the experimental IFT measurements appears to be excellent in simple binary systems. This has been found to be true even in our measurements using the n-decane-CO_2 binary system at 37.8°C and at pressures ranging from 0.1 to 7.7 MPa, for which the results are shown in Fig. 1. However, parachor model results are reported to be in poor agreement with measured IFT data when extended to complex multicomponent hydrocarbon mixtures [15]. Our previously published work [7] also shows a poor match of parachor model predictions with IFT measurements in a quaternary system consisting of live decane, which was a mixture of 25 mole% n-C_1, 30 mole% n-C_4, and 45 mole% n-C_{10}, and CO_2 gas at 71.1°C and within the pressure range of 7.7–12.2 MPa. These results are shown in Fig. 2.

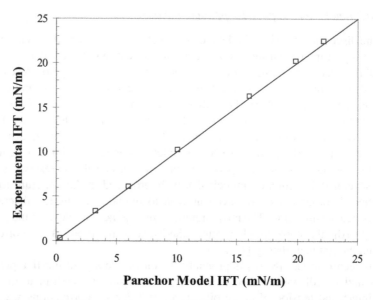

Figure 1. Experimental IFT vs. parachor model predictions in n-decane-CO_2 binary system at 37.8°C (data from Ref: 7).

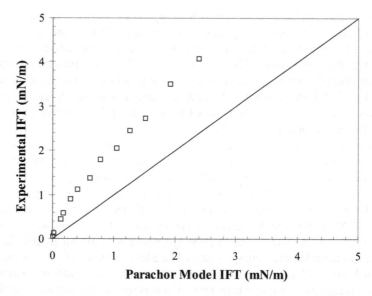

Figure 2. Experimental IFT vs. parachor model predictions in live decane (25 mole% n-C_1, 30 mole% n-C_4, and 45 mole% n-C_{10})-CO_2 quaternary system at 71.1°C (data from Ref: 7). Note: The straight line in the above figure is a 45° line and the proximity of the points to this line in the plot indicates a good match between the experimental IFT data and parachor model IFT predictions.

The poor performance of parachor model in multicomponent systems appears to be mainly due to using mole fraction weighted empirical mixing rules to obtain the parachors of the mixtures from pure component values. Several attempts have been made in the literature to improve the parachor model IFT predictions in multicomponent hydrocarbon systems. Fawcett [16] has reviewed these studies in detail. Most of these studies have focused on matching experimental IFT data in multicomponent systems using modified empirical mixing rules and hence there appears to be no strong theoretical background associated with them.

The corresponding states theory [11] correlates IFT as a unique function of reduced temperature and is only applicable to pure fluids. The corresponding states theory models cannot be extended to multicomponent mixtures as the compositions of vapor and liquid change with pressure and temperature.

Eckert and Prausnitz [12] proposed the basic thermodynamic model to determine the surface tension of pure liquids using the thermodynamic properties of molecules such as cell partition functions, configurational molar energies, and chemical potentials. They later extended their thermodynamic model to compute the interfacial tension of binary liquid mixtures and the model predictions were shown to be in good agreement with the experimental data. However, the application of thermodynamic models to test the measured IFT data in hydrocarbon systems, such as crude oils, containing multiple components in both the vapor and liquid phases has not been documented.

The gradient theory [13] requires mainly two parameters, the free-energy density of the bulk fluid and the influence parameter of the interface, to calculate the interfacial tension between the fluid phases. The free-energy density of the bulk homogeneous fluid is determined using a suitable equation of state. The influence parameter carries the information on the molecular structure of the interface and it essentially determines the density gradient with response to the local deviation of chemical potential from its corresponding bulk phase value. The influence parameter is determined using the empirical mathematical equations and this parameter together with a homogeneous fluid equation of state is used to characterize the non-homogeneous vapor-liquid interface in the gradient theory. The mathematical expressions proposed in the literature for influence parameter calculations, by fitting the experimental surface tension data of several pure fluids, are not fully satisfactory. Miqueu *et al.* [17] thoroughly discussed the deficiencies of various influence parameter equations. The gradient theory has been successfully used in the recent literature to determine the interfacial tension of pure fluids and binary mixtures. However, due to the deficiencies in the influence parameter correlations, and elaborative calculation procedures and lack of improved results relative to the other IFT predictive models, this method has not received much attention in the petroleum industry.

1.3. New mechanistic parachor IFT model

Recently, a modified version of the parachor model called the "mechanistic parachor model" has been proposed [18] to determine the interfacial tension in crude oil-solvent systems and to overcome the inabilities of the traditional parachor model in multicomponent hydrocarbon systems. While applying the traditional parachor model to multicomponent mixtures, parachor values of pure components are used in IFT predictions, considering each component of the mixture as if all the others were absent. Due to this assumption, the interactions of each component with the others in a multicomponent mixture are neglected, which appears to be the main reason for poor IFT predictions from the parachor model when applied to multicomponent hydrocarbon systems.

Hence, the ratio of diffusivity coefficients between the fluid phases raised to an exponent is introduced into the parachor model as a correction factor for pure component parachors to account for various mass transfer interactions taking place within a multicomponent mixture. The resulting model after incorporating mass transfer effects is called as the "mechanistic parachor model" due to its ability to characterize the mass transfer mechanisms (vaporization or condensation) responsible for attaining fluid phase equilibria. The mass transfer-corrected parachor value of a component in a mixture is given by,

$$P_i^{cor} = P_i \left(\frac{D_{l-v}}{D_{v-l}} \right)^n \tag{2}$$

The mechanistic parachor model is expressed by,

$$\sigma^{1/4} = \rho_M^L \sum x_i P_i^{cor} - \rho_M^V \sum y_i P_i^{cor} \tag{3}$$

Combining Eqs. (2) and (3) yields,

$$\sigma^{1/4} = \left(\frac{D_{l-v}}{D_{v-l}} \right)^n \left(\rho_M^L \sum x_i P_i - \rho_M^V \sum y_i P_i \right) \tag{4}$$

where P_i^{cor} is the mass transfer-corrected parachor value of a compound in a mixture, P_i is the parachor value of pure compound, D_{l-v} is the diffusivity of liquid phase in vapor, D_{v-l} is the diffusivity of vapor phase in liquid and n is the exponent, whose sign and value characterize the type and extent of governing mass transfer mechanism responsible for fluid phase equilibria. The value of n greater than zero ($n > 0$) indicates the governing mechanism to be vaporization of lighter components from the oil to the gas phase. The governing mechanism for fluid phase equilibria is condensation of intermediate to heavy components from the gas to the crude oil for n values less than zero ($n < 0$). The value of n equal to zero ($n \approx 0$) indicates equal influence of vaporization and condensation mass transfer mechanisms in attaining fluid phase equilibria. This condition of equal mass transfer in both vaporization and condensation directions appears to be the most common in binary mixtures where the traditional parachor model showed reasonably accurate interfacial tension predictions ($n = 0$ in the mechanistic parachor model). The higher the numerical value of n (irrespective of its sign), the greater is the extent of the corresponding governing mass transfer mechanism.

The Wilke and Chang [19] empirical correlation has been used to compute the diffusivity coefficients between the vapor and liquid phases, as given by,

$$D_{AB} = \frac{(117.3 \times 10^{-18})(\varphi M_B)^{0.5} T}{\mu v_A^{0.6}} \tag{5}$$

where D_{AB} is diffusivity of solute A in very dilute solution in solvent B (m²/s), M_B is the molecular weight of the solvent (kg/kmol), T is the temperature (K), μ is the solution viscosity (kg/m.s or Pa.s), v_A is the solute molal volume at normal boiling point (m³/kmol), and φ is the association factor for solvent (set equal to unity since the fluids used in this study are unassociated).

The Wilke and Chang correlation was extended to multicomponent mixtures using the mole fraction weighted averaging techniques. The values of molar volumes at normal boiling point for different components needed for vapor-liquid diffusivity calculations have been obtained from reference [20].

The exponent (n) in the mechanistic model is computed using the mass transfer enhancement parameter (k) and the ratio of diffusivity coefficients between the fluid phases, as given by:

$$k = \left(\frac{D_{l-v}}{D_{v-l}} \right)^n \tag{6}$$

The mass transfer enhancement parameter (k) is independently determined as the correction factor to the original parachor model at which the objective function (Δ, the sum of weighted squared deviations between the original parachor model predictions and the experimental IFT values) becomes the minimum.

$$\Delta = \sum_{j=1}^{N} \left[w_j \left(\frac{\sigma_j^{pred}(X) - \sigma_j^{exp}}{\sigma_j^{exp}} \right) \right]^2 \tag{7}$$

Here σ^{pred} is the predicted IFT value from the parachor model, σ^{exp} is the experimental IFT value, w is the weighting factor (always equal to unity in this study), N is the number of measured IFT data points to be fitted, and X designates the correction factor, which is equal to the ratio of experimental IFT value to the original parachor model prediction.

The use of diffusivities in the mechanistic parachor model and its ability to provide information on mass transfer mechanisms signifies the dynamic nature of the model. Crude oils contain thousands of chemical components and hence it is difficult to attain the thermodynamic equilibrium compositions of these large numbers of components within finite aging periods between the fluid phases. However, after a certain aging period, the changes in IFT with time become so minute that it is reasonable to approximate them as near-equilibrium interfacial tensions. Ayirala

S. C. Ayirala and D. N. Rao

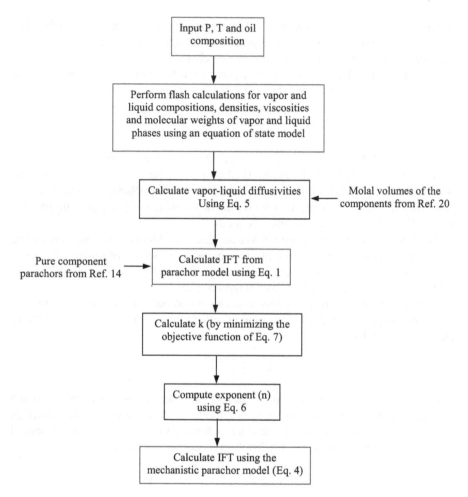

Figure 3. Flowchart showing the step-wise procedure used for IFT calculations in the new mechanistic parachor model.

and Rao [21] measured dynamic behavior of IFT in simple standard gas-oil systems and observed changes in IFT until up to the aging period of about 50 hours between the fluid phases. They also reported that the one-hour aging period between the fluid phases had accounted for nearly 99.5–99.7% of the equilibrium IFT value. Hence there would always be some dynamic behavior associated with vapor-liquid interfacial tension of crude oils, which, therefore, justifies the use of the mechanistic parachor model involving diffusivities for vapor-liquid IFT calculations. The flowchart that describes the step-wise procedure used for the mechanistic parachor model IFT calculations is shown in Fig. 3.

Table 1.
Physical properties of crude oils used (Firoozabadi *et al.* [22])

Crude oil	Saturation pressure (MPa)	Temperature (°C)	(C_1-C_3) Mole%	C_{7+} Mol. wt	C_{7+} Sp. gravity
A	14.9	54.4	46.5	227.4	0.870
C	31.6	82.2	64.6	217.0	0.838
D	17.7	76.7	51.2	234.3	0.868

2. OBJECTIVES

The objectives of this study were to utilize the newly proposed mechanistic parachor model to compute the vapor-liquid interfacial tension of crude oils and to determine the governing mass transfer mechanism responsible for attaining vapor-liquid phase equilibria in these crude oil systems. For this purpose, three reservoir crude oil systems (A, C, and D) were chosen from the literature, for which the phase behavior data required for IFT calculations and the experimental vapor-liquid interfacial tensions were readily available [22]. The reported experimental IFT measurements were obtained using the pendent drop technique. Vapor-liquid interfacial tension calculations based on the parachor model were carried out using the QNSS/Newton flash algorithm [23] and the Peng-Robinson equation of state [24], available within a commercial software package [25]. Diffusivities were computed using the flashed individual fluid phase compositions and the fluid phase viscosities obtained from the Pederson's corresponding states model [26]. The three crude oils used for IFT modeling in this study were from different reservoirs with specific gravities ranging from 31 to 35° API (0.87 to 0.85 g/cm^3). The key physical characteristics of the crude oils used are described in Table 1. From Table 1, it can be seen that crude oil C is the lightest when compared to the other two crude oils A and D, as it has higher C_1-C_3 molar composition and lower C_{7+} molecular weight.

3. RESULTS AND DISCUSSION

3.1. Reservoir crude oil A

The comparison of original parachor model IFT predictions with experiments at various pressures and at a reservoir temperature of 54.4°C is given in Table 2 for this crude oil system. These results are also shown in Fig. 4. As can be seen from Fig. 4, significant IFT under-predictions are obtained with the parachor model when compared to the experiments. The absolute deviations as high as 45% were obtained between the experimental values and the parachor model predictions.

The disagreement between the experiments and the model predictions, as seen in Fig. 4, can be attributed mainly to the absence of mass transfer effects in the original parachor model. These mass transfer effects refer to the evaporation of lighter components (C_1-C_3) from crude oil into the vapor phase as the pressure

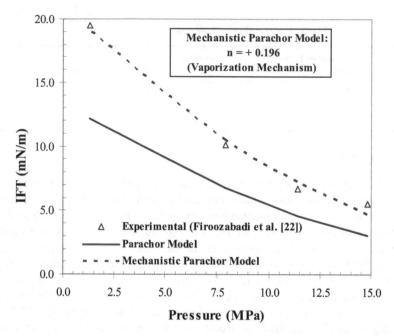

Figure 4. Comparison of IFT measurements with parachor and mechanistic parachor model predictions for crude oil A at 54.4°C.

Table 2.
Comparison of IFT measurements with parachor and mechanistic parachor model predictions for crude oil A at 54.4°C

Pressure (MPa)	Expl. IFT (mN/m)	Parachor model		Mechanistic parachor model	
		IFT (mN/m)	Abs. dev. (%)	IFT (mN/m)	Abs. dev. (%)
14.8	5.5	3.0	44.9	4.8	13.5
11.4	6.7	4.6	31.2	7.2	8.0
7.9	10.1	6.7	33.4	10.6	4.6
1.3	19.5	12.2	37.6	19.1	2.1
Average			36.8		7.1

declines below the saturation pressure. The lower the pressure, the higher the extent of evaporation and the larger the compositional changes in the vapor and liquid phases. This is reflected in the larger deviations observed in the parachor model predictions from the experimental IFT values at lower pressures. This confirms the absence of mass transfer effects as being the main cause for the deviations in the parachor model predictions.

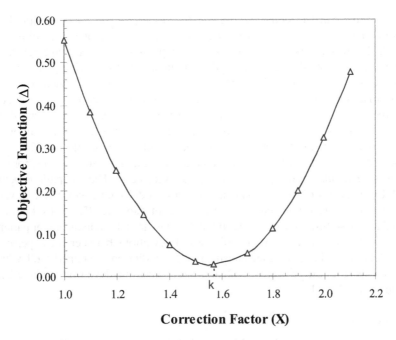

Figure 5. Minimization of objective error function (Δ) using various parachor model correction factors (X) to determine the mass transfer enhancement parameter (k) for crude oil system 'A'.

Table 3.
Vapor-liquid diffusivities for crude oil A at 54.4°C

Pressure (MPa)	D_{l-v} (m^2/s)	D_{v-l} (m^2/s)	D_{l-v}/D_{v-l}
14.8	2.5E-08	3.9E-09	6.4
11.4	2.7E-08	3.6E-09	7.4
7.9	2.8E-08	3.2E-09	8.8
1.3	3.0E-08	1.7E-09	17.4
Average			10.0

Hence correction factors are used for the original parachor model predictions to minimize the objective function (Δ), which is the sum of weighted squared deviations between the model predictions and experimental values. The correction factors, X (as in Eq. 7) and the resulting objective functions for this crude oil system are shown in Fig. 5, as an example case. The mass transfer enhancement parameter (k), which is the correction factor at which objective function becomes the minimum, is found to be 1.57 for crude oil A in Fig. 5.

The computed diffusivities between the vapor and liquid phases at various pressures are given in Table 3. Interestingly, there is not much change in the ratio of diffusivities in both directions (liquid to vapor and vapor to liquid) with pressure. The average ratio of diffusivities between the fluid phases at all pressures is 10.0. From the mass transfer enhancement parameter and the average ratio of diffusivities between the fluid phases, the exponent (n) characterizing the governing mass transfer mechanism is found from Eq. 6 to be $+0.196$. This value of n, being greater than zero, indicates that the vaporization of light components from the crude oil is the mass transfer mechanism that governs the fluid phase equilibria of vapor and liquid phases. This can be attributed to the presence of significant amounts of lighter components (46.5 mole% C_1 to C_3) in this particular crude oil [22].

The above value of n ($+0.196$) was then used in the mechanistic parachor model (Eq. 4) to recalculate the vapor-liquid interfacial tensions. The comparison between the IFT predictions of the mechanistic parachor model with experiments at various pressures and at reservoir temperature is given in Table 2. These results are also plotted in Fig. 4. Since the optimization of the mass transfer enhancement parameter (k) is based on minimizing the sum of squared deviations between the experimental and calculated values, the mechanistic model predictions matched well with the experiments at all the pressures (with an average absolute deviation of about 7%) as can be seen in Fig. 4.

3.2. Reservoir crude oil C

The original parachor model predictions and the experimental IFT values for this reservoir crude oil at different pressures and at the reservoir temperature of 82.2°C are shown in Table 4 and Fig. 6. Poor match between the parachor model predictions and experimental IFT values can be seen with an average absolute deviation of 57%. The minimization of the objective function (Δ) yielded a mass transfer enhancement parameter (k) of 2.32. The calculated diffusivities between the vapor and liquid phases of this crude oil at various pressures are shown in Table 5. From Table 5, it can be seen that the ratio of diffusivities between the fluid phases remains nearly constant irrespective of the pressure, and the average ratio of diffusivities between the fluid phases at all pressures is determined to be 3.57. A mechanistic model exponent (n) of $+0.662$ was obtained and the positive value of the exponent

Table 4.

Comparison of IFT measurements with parachor and mechanistic parachor model predictions for crude oil C at 82.2°C

Pressure (MPa)	Expl. IFT (mN/m)	Parachor model		Mechanistic parachor model	
		IFT (mN/m)	Abs. dev. (%)	IFT (mN/m)	Abs. dev. (%)
26.3	1.3	0.5	58.6	1.2	4.0
22.9	2.3	0.9	61.5	2.1	10.6
19.4	3.3	1.4	57.0	3.3	0.2
16.0	4.6	2.2	51.5	5.2	12.5
Average			57.1		6.8

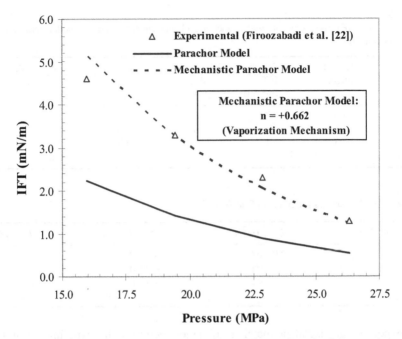

Figure 6. Comparison of IFT measurements with parachor and mechanistic parachor model predictions for crude oil C at 82.2°C.

Table 5.
Vapor-liquid diffusivities for crude oil C at 82.2°C

Pressure (MPa)	D_{l-v} (m²/s)	D_{v-l} (m²/s)	D_{l-v}/D_{v-l}
26.3	2.5E-08	8.4E-09	2.9
22.9	2.6E-08	7.8E-09	3.3
19.4	2.7E-08	7.2E-09	3.8
16.0	2.9E-08	6.6E-09	4.3
Average			3.6

obtained once again indicates that vaporization of lighter components from the crude oil is mainly responsible for vapor-liquid phase equilibria.

The higher value of n (+0.662) obtained for this crude oil system compared to reservoir crude oil A (+0.196) implies more pronounced vaporization mass transfer effects in reservoir crude oil C. This can be attributed to the presence of relatively larger amounts of lighter components (64.6 mole% C_1 to C_3) in the reservoir crude oil C when compared to 46.5 mole% C_1 to C_3 in reservoir crude oil A. The comparison between the mechanistic parachor model IFT predictions and the experiments at various pressures is given in Table 4 and shown in Fig. 6.

Table 6.
Comparison of IFT measurements with parachor and mechanistic parachor model predictions for crude oil D at 76.7°C

Pressure (MPa)	Expl. IFT (mN/m)	Parachor model		Mechanistic parachor model	
		IFT (mN/m)	Abs. dev. (%)	IFT (mN/m)	Abs. dev. (%)
13.9	6.0	2.7	54.5	5.3	12.3
11.1	8.5	4.2	51.1	8.0	5.8
7.7	10.3	6.1	41.0	11.7	13.6
Average			48.8		10.6

Table 7.
Vapor-liquid diffusivities for crude oil D at 76.7°C

Pressure (MPa)	D_{l-v} (m^2/s)	D_{v-l} (m^2/s)	D_{l-v}/D_{v-l}
13.9	2.9E-08	5.1E-09	5.5
11.1	3.0E-08	4.9E-09	6.1
7.7	3.1E-08	4.5E-09	6.9
Average			6.2

As expected, an excellent match is obtained between the experiments and the mechanistic model predictions with an average absolute deviation less than 6.8%.

3.3. Reservoir crude oil D

The comparison between the vapor-liquid interfacial tensions from the parachor model and the experiments for this crude oil system at 76.7°C are given in Table 6 and shown in Fig. 7. Once again, poor performance of the parachor model can be seen for this crude oil system also with absolute deviations in the range of 41–55%, when compared to the experiments. A mass transfer enhancement parameter (k) of 1.88 was obtained in this case by minimizing the sum of weighted squared deviations between the parachor model predictions and experimental IFT values. The resultant vapor-liquid diffusivities obtained from the Wilke and Chang [19] correlation are given in Table 7. The ratio of vapor-liquid diffusivities remained almost unchanged with pressure. These findings are similar to those observed with the other two crude oil systems. The average ratio of diffusivities between the fluid phases at various pressures is computed to be 6.19. From the mass transfer enhancement parameter and the average ratio of diffusivities between the fluid phases, the exponent (n) in the mechanistic model is calculated as +0.346. The positive sign of n indicates that even for this crude oil system, vaporization of light components from the crude oil is the dominating mass transfer mechanism for attaining the vapor-liquid phase equilibria.

The value of n (+0.346) obtained for this crude oil system is relatively high when compared to crude oil A (+0.196) and is relatively low when compared to the crude

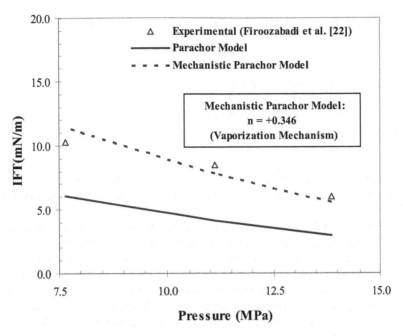

Figure 7. Comparison of IFT measurements with parachor and mechanistic parachor model predictions for crude oil D at 76.7°C.

oil C (+0.662). This indicates relatively higher vaporization mass transfer effects in crude oil D compared to crude oil A, but relatively lower vaporization mass transfer effects compared to the crude oil C. This can be attributed to the presence of relatively larger amounts of lighter components (51.2 mole% C_1 to C_3) in the reservoir crude oil D compared to reservoir crude oil A (46.5 mole% C_1 to C_3) and due to relatively smaller amounts of lighter components compared to the reservoir crude oil C (64.6 mole% C_1 to C_3). The comparison of the experimental IFT measurements with the mechanistic parachor model predictions for this particular crude oil system is given in Table 6 and shown in Fig. 7. As before with the other two crude oil systems, good match between the mechanistic parachor model predictions and the experimental IFT data can be seen even in this case with an average absolute deviation of 10.6%.

3.4. Parachor physics and thermodynamics

A detailed literature review was conducted to understand the term 'parachor' so as to conceive the underlying physics and thermodynamics behind the 'parachor' and the results of this literature study are provided in this section.

Exner [27] defined parachor as the molar volume at such a temperature at which surface tension has the unit value as long as this temperature does not approach the critical temperature, as described by the following Eqs. (8) and (9).

$$P = \frac{M}{\rho_l - \rho_v} \sigma^{1/4} \qquad (8)$$

where P is the parachor, ρ_l and ρ_v denote densities of liquid and vapor phases, respectively, σ is the surface tension, and M is the molecular weight.

At temperatures lower than the critical temperature, ρ_v can be neglected when compared to ρ_l and hence Eq. (8) simplifies to,

$$P = M\rho_l^{-1}\sigma^{1/4} \qquad (9)$$

Parachor is compound specific. Parachor is temperature independent at temperatures below the critical temperature for all non-polar and slightly polar compounds [27].

Parachor value of a compound is related to its molecular weight. Firoozibadi et $al.$ [22] used the data from Katz et $al.$ [28] and Rossini [29] to show a linear straight-line relationship between the parachor and molecular weight for n-alkanes. They also computed the parachors of several distillation cuts of various crude oils from surface tension measurements and showed a quadratic relationship between the parachor and molecular weight for all the crude cuts, except for the residues. They attributed this discontinuity for the last heavy residue fractions largely to the presence of asphaltene materials in them.

Parachor value of a compound does not depend on pressure [22]. Firoozabadi et $al.$ [22] determined the parachors of different crude cuts of various crude oils at different pressures and reported similar parachor values for individual crude cuts at all the pressures tested.

Parachor value of a mixture is related to solute concentration [30, 31]. Hammick and Andrew [30] computed parachor values of mixtures of benzene (non-associated solvent) with various non-associated solutes such as carbon tetrachloride, m-xylene, cyclohexane, and chloroform, using surface tension measurements. They found that the parachor values of the solution were linearly related to solute concentration and either increased or decreased as the solute concentration in the solution was increased. The data from the study of Hammick and Andrew [30] are plotted in Fig. 8.

3.5. Relationship between the parachor and the exponent (n) in the mechanistic parachor model

Interestingly, the exponent (n) in the mechanistic parachor model exhibits similar characteristics as the parachor. The exponent is specific for a crude oil (liquid and vapor phases). It is independent of pressure (as can be seen in Figs. 4, 6 and 7, where a single value of mechanistic parachor model exponent was able to accurately match IFT values at all pressures for the respective crude oil system considered). The exponent also appears to be independent of temperature, although this needs to be still verified. Based on these observations, like for the parachor,

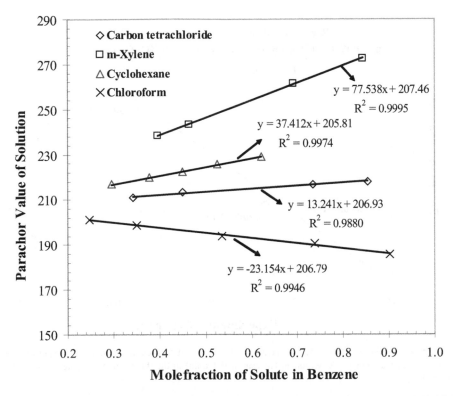

Figure 8. Effect of solute composition in benzene on parachor value of solution (data from Ref: 30).

a linear relationship between the exponent and solute composition is hypothesized. For testing this hypothesis, the crude oil systems (A, C and D) of this study were utilized.

In crude oil systems such as crude oils A, C and D, the vapor phase is formed primarily due to vaporization of lighter components (C_1-C_3) from the crude oil. Therefore, the composition of lighter ends (C_1-C_3) in the crude oil constitutes the solute composition. CO_2 in crude oil has also been included, as it can be considered as an active component involved in the vaporization process. Hence the mechanistic parachor model exponents for these three crude oil systems are related to the normalized solute composition $(C_1-C_3+CO_2)/(C_4-C_{/+})$ in the crude oil, using regression analysis. A good linear relationship between the exponent and the normalized solute composition was observed with a determination coefficient (R^2) of 0.965 and the resulting regression equation obtained is given as:

$$n = -0.1216 + 0.3950 \left(\frac{Mole\%CO_2 + C_1 - C_3}{Mole\%C_4 - C_{7+}} \right)_{Oil} \tag{10}$$

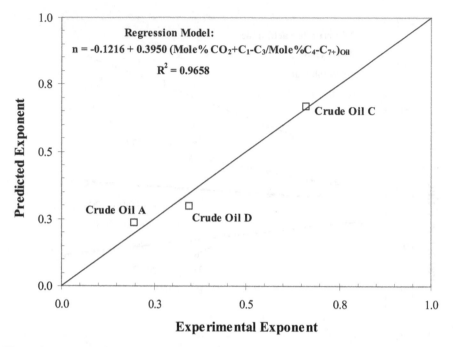

Figure 9. Predicted mechanistic parachor model exponent (regression model) vs. experimental mechanistic parachor model exponent (measured IFT data) in the three crude oil systems of "A, C, and D".

The comparison of predicted mechanistic parachor model exponent using the regression Eq. (10) with the experimental mechanistic parachor model exponent determined using the measured IFT data in the three crude oil systems of A, C, and D is shown Fig. 9. Good match between the predicted and experimental mechanistic parachor model exponents can be seen in Fig. 9 with all the points falling close to the 45° line. Hence, the regression Eq. (10) can be used for *a priori* prediction of exponent (n) in the mechanistic parachor model for crude oil systems simply by knowing the composition of crude oil, without fitting any experimental data.

The summary of similarities observed in the characteristics between the mechanistic parachor model exponent (n) and the parachor are shown in Table 8. From Table 8, it can be seen that parachor and the exponent (n) in the mechanistic parachor model have similar characteristics. Thus this interesting feature observed during the course of this study has been well utilized to develop a generalized regression model for the mechanistic parachor model exponent prediction by simply using the compositional data of reservoir fluid and thereby eliminating the need for any experimental IFT data.

Table 8.
Summary of similarities observed between the parachor and the mechanistic parachor model exponent (n)

Parachor	Mechanistic parachor model exponent (n)
Compound specific	Specific for a crude oil
Independent of temperature	Appears to be temperature independent and still needs to be examined
Independent of pressure	Independent of pressure
Linearly related to solute concentration	Linearly related to solute composition present in the crude oil

3.6. Application of the proposed regression model to a new crude oil system

The proposed generalized regression model was utilized to predict the exponent in the mechanistic parachor model and, consequently, vapor-liquid interfacial tensions in the Prudhoe Bay crude oil system for validation. This crude oil has 47.41 mole% C_1-C_3, 43.99 mole% C_4-C_{7+} and 8.42 mole% CO_2 [32]. The vapor-liquid experimental IFT data on the Prudhoe Bay crude oil measured using surface laser light scattering spectroscopy at 93.3°C from Dorshow [33] were used for comparison with the results from the proposed regression model. The IFT measurements reported only in the range of pressures 1.8–7.0 MPa were used, as the measurements at pressures higher than 7.0 MPa appear to be affected by asphaltene precipitation (as evidenced by the different rates of decline in vapor-liquid interfacial tension). The phase behavior calculations, to generate the input data required for mechanistic parachor model (such as equilibrium vapor and liquid compositions, and the properties of fluid phases), were carried out using the Soave-Redlich-Kwong equation of state [34, 35] for this new crude oil system. The measured liquid phase viscosities from reference [33] were used during the vapor-liquid diffusivity calculations.

A mechanistic parachor model exponent of 0.380 was obtained by using the compositional data of reservoir crude oil in the proposed generalized regression model. This exponent computed using the regression model deviated by only about 5.9% from the mechanistic parachor model exponent of 0.404 obtained by fitting the available vapor-liquid IFT experimental data.

The comparison of the IFT measurements with the predictions of the parachor model and the mechanistic parachor model with the exponents calculated using the crude oil compositional data as well as all the available IFT experimental data is shown in Fig. 10. From Fig. 10, significant IFT under-predictions can be seen with the original parachor model when compared to the experiments. However, an excellent match of IFT predictions with experiments is obtained with both mechanistic parachor models (one with the exponent obtained by fitting all the experimental IFT data and the other with the exponent calculated using crude oil compositional data in Eq. 10). Furthermore, almost similar IFT predictions can be

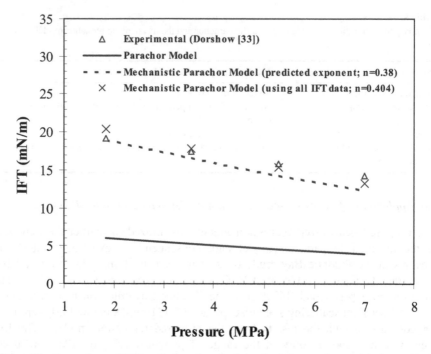

Figure 10. Comparison of IFT predictions from both mechanistic parachor models (one with the exponent calculated using Eq. 10 and the other using the exponent obtained by fitting all the experimental IFT data) with measured IFT data in Prudhoe Bay crude oil system.

seen from the mechanistic parachor model for both these exponents used. Thus, this validates the proposed regression model to predict the exponent in the mechanistic parachor model without fitting any experimental IFT data.

4. SUMMARY AND CONCLUSIONS

A new mechanistic parachor model has been developed and utilized to predict vapor-liquid interfacial tension for three different crude oil systems. The mechanistic parachor model was able to accurately predict vapor-liquid interfacial tensions in all three crude oil systems studied, with average absolute deviations less than 10%. The overall diffusivity ratio between the fluid phases raised to an exponent seems to be adequate to account for the interaction of each individual component with the rest in a mixture and this appears to be the main reason for the good IFT predictions from the mechanistic parachor model. The positive exponents obtained in the mechanistic parachor model for all the three crude oils indicate that vaporization of lighter components from the crude oil is the governing mass transfer mechanism responsible for attaining vapor-liquid phase equilibria. The mechanistic parachor model exponent was found to have similar characteristics as the parachor. A generalized regression model was developed by correlating the exponent in the mechanistic parachor model with the normalized solute composition present in the crude oil. This regression model has been validated for predicting the mechanistic parachor model exponent using the Prudhoe Bay crude oil and hence can be used for *a priori* prediction of exponent in the mechanistic parachor model simply by using the compositional data of crude oil, without the need of any experimental IFT data.

Acknowledgements

The work reported in this paper is part of a project supported financially by the U.S Department of Energy under Award No. DE-FC26-02NT-15323. Any opinions, findings, conclusions or recommendations expressed herein are those of authors and do not necessarily reflect the views of the US-DOE. The financial support of this project by the U.S. Department of Energy is gratefully acknowledged. The authors thank Dr. Jerry Casteel and Dr. Betty Felber of NPTO/DOE for their support and encouragement. Sincere thanks are also due to Daryl S. Sequeira and Wei Xu of Louisiana State University for their immense help in the project.

REFERENCES

1. D. N. Rao and S. C. Ayirala, *J. Adhesion Sci. Technol.* **20**, 125 (2006).
2. A. L. Benham, W. E. Dowden and W. J. Kunzman, *J. Petroleum Technol.* **12**, 229 (1960).
3. L. W. Holm, in: *Miscible Displacement*, H. B. Bradley (Ed.), pp 1-45, Society of Petroleum Engineers, Richardson, TX (1987).
4. L. W. Lake, *Enhanced Oil Recovery*, p. 234, Prentice-Hall, Englewood Cliffs, NJ (1989).
5. D. N. Rao, *Fluid Phase Equilibria* **139**, 311 (1997).
6. D. N. Rao and J. I. Lee, *J. Petroleum Sci. Eng.* **35**, 247 (2002).
7. S. C. Ayirala, W. Xu and D. N. Rao, *Canadian J. Chem. Eng.* **84**, 22 (2006).
8. D. B. Macleod, *Trans. Faraday Soc.* **19**, 38-42 (1923).
9. S. Sugden, *J. Chem. Soc.* **125**, 32-41 (1924).
10. C. F. Weinaug and D. L. Katz, *Ind. Eng. Chem.* **35**, 239-246 (1943).
11. H. L. Brock and R. B. Bird, *AIChE J.* **1**, 174 (1955).
12. C. A. Eckert and J. M. Prausnitz, *AIChE J.* **10**, 677 (1964).
13. B. S. Carey, L. E. Scriven and H. T. Davis, *J. Chem. Phys.* **69**, 5040 (1978).
14. A. Danesh, *PVT and Phase Behavior of Petroleum Reservoir Fluids*, pp. 281-299, Elsevier, Amsterdam (1998).
15. A. S. Danesh, A. Y. Dandekar, A. C. Todd and R. Sarkar, Paper No. SPE 22710, *Proc. 66ᵗʰ SPE Annual Technical Conference and Exhibition*, Dallas, TX (1991).
16. M. J. Fawcett, Paper No. SPE 28611, *Proc. 69ᵗʰ SPE Annual Technical Conference and Exhibition*, New Orleans, LA (1994).
17. C. Miqueu, B. Mendiboure, A. Graciaa and J. Lachaise, *Fluid Phase Equilibria* **207**, 225 (2003).
18. S. C. Ayirala and D. N. Rao, *J. Colloid Interface Sci.* **299**, 321 (2006).
19. C. R. Wilke and P. Chang, *AIChE J.* **1**, 264 (1955).
20. R. E. Treybal, *Mass-Transfer Operations*, p. 33, McGraw-Hill Book Company, Singapore (1981).
21. S. C. Ayirala and D. N. Rao, Paper No. SPE 99606, *Proc. SPE/DOE Symposium on Improved Oil Recovery*, Tulsa, OK (2006).

22. A. Firoozabadi, D. L. Katz, H. Saroosh and V. A. Sajjadian, *SPE Reservoir Eng. J.* **3**, 265 (1988).
23. L. X. Nghiem and R. A. Heidemann, *Proc. 2nd European Symposium on Enhanced Oil Recovery*, Paris, France (1982).
24. D. Y. Peng and D. B. Robinson, *Ind. Eng. Chem. Fundam.* **15**, 59 (1976).
25. Winprop Phase Property Program, Version 2001 User's Guide, Computer Modeling Group Ltd., Calgary, Alberta, Canada (2001).
26. K. S. Pederson and A. A. Fredenslund, *Chem. Eng. Sci.* **42**, 182 (1987).
27. O. Exner, *Collect. Czech. Chem. Commun.* **32**, 24 (1967).
28. D. L. Katz, R. R. Monroe and R. P. Trainer, *AIME Tech. Publ.* No. 1624 (1943).
29. F. D. Rossini, *Selected Values of Physical and Thermodynamic Properties of Hydrocarbons and Related Compounds*, Carnegie Press, Pittsburgh (1953).
30. D. L. Hammick and L. W. Andrew, *J. Chem. Soc.* **130**, 754 (1929).
31. S. T. Bowden and E. T. Butler, *J. Chem. Soc.* **140**, 79 (1939).
32. A. P. Spence, J. F. Ostrander, Paper No. SPE 11962, *Proc. SPE 58th Annual Technical Conference and Exhibition*, San Francisco, CA (1983).
33. R. B. Dorshow, *SPE Adv. Technology Series* **3**, 120 (1995).
34. O. Redlich and J. N. S. Kwong, *Chem. Rev.* **44**, 233 (1949).
35. G. Soave, *Chem. Eng. Sci.* **27**, 1197 (1972).

Contact Angle, Wettability and Adhesion, Vol. 5, pp. 95–111
Ed. K.L. Mittal
© VSP 2008

Influence of ambient humidity on the apparent surface free energy of poly(methyl methacrylate) (PMMA)

LUCYNA HOLYSZ, EMIL CHIBOWSKI* and KONRAD TERPILOWSKI

*Department of Physical Chemistry-Interfacial Phenomena, Faculty of Chemistry,
Maria Curie-Sklodowska University, 20-031 Lublin, Poland*

Abstract—The total surface free energy of poly(methyl methacrylate) (PMMA) was determined from the measured advancing and receding contact angles of water for at least 20 droplets, at varying relative humidity (RH) of the atmosphere present in the measuring chamber. At three selected humidities contact angles of diiodomethane and formamide were also determined. For determination of total surface free energy and its components two approaches were used: the contact angle hysteresis (CAH) approach and the Lifshitz-van der Waals acid-base (LWAB) approach. In the former approach the advancing and receding contact angles are employed, and in latter only the advancing angle. It was found that the apparent total surface free energy γ_S^{tot} of PMMA in the RH range 35–100% decreases roughly linearly and in the range RH 1.5–10% it is practically constant. However, at RH between 10% and 35% a sharp minimum appears at RH = 15%. Possible reasons for such behaviour are discussed and a mechanism based on the coalescence of pre-adsorbed water nano- and/or micro-droplets during their contact with the water droplet deposited on the surface is proposed.

Keywords: PMMA; contact angle hysteresis; relative humidity; surface free energy.

1. INTRODUCTION

In various physicochemical processes taking place both in natural environment and industry wetting and the surface free energies of solid and liquid, and the solid/liquid interfacial free energy play a crucial role. However, both the thermodynamic description and experimental determination of solid surface free energy are still an open problem. The magnitude of the surface free energy results from the kind and strength of intermolecular interactions. Contact angle is an important parameter in wetting processes, because its measurement allows evaluation of solid surface free energy. The fundamental equation which is used for this purpose is still Young's equation [1]:

$$\gamma_S = \gamma_{SL} + \gamma_L \cos\theta \tag{1}$$

*To whom correspondence should be addressed. Tel.: (48-81) 537-5651; Fax: (48-81) 533-3348;
e-mail: emil@hermes.umcs.lublin.pl

where: γ_S is the solid surface free energy, γ_L is the liquid surface tension, γ_{SL} is the solid-liquid interfacial free energy, and θ is the contact angle.

Equation (1) thus written means that the solid surface is bare and the contact angle corresponds to the advancing contact angle θ_a:

$$\gamma_S = \gamma_{SL} + \gamma_L \cos\theta_a \tag{2}$$

In the case where the liquid vapour has adsorbed behind the droplet, the equation reads:

$$\gamma_{Sf} = \gamma_{SL} + \gamma_L \cos\theta_r \tag{3}$$

where: γ_{Sf} is the surface free energy of the solid on which the liquid film is present (the film-covered solid surface free energy), and θ_r is the receding contact angle. The γ_{Sf} is expressed as:

$$\gamma_{Sf} = \gamma_S + \pi \tag{4}$$

where π is the film pressure.

When the liquid does not spread completely on the solid surface (so-called contact-angle liquid) the film pressure is positive and the film should increase the apparent surface free energy of the solid surface. In case of a spreading liquid (which does not form a definite contact angle) when the film forms, as a result of the surface pre-wetting or the vapor adsorption, the film pressure is negative and such film decreases the solid surface free energy [1].

In the Young equation contact angle and liquid surface tension are measurable quantities, but the solid surface free energy and the solid/liquid interfacial free energy are unknown and, therefore, this equation cannot be solved. However, the work of adhesion W_A can be determined [1] as:

$$W_A = \gamma_L(1 + \cos\theta) \tag{5}$$

For a solid, its surface free energy can be determined if the work of adhesion is formulated in such a way that it involves the solid surface free energy, but this problem is not fully solved yet. Intermolecular forces, i.e. dispersion, dipole-dipole, π-electrons, hydrogen bonding, or generally Lewis acid-base, electron-donor and electron-acceptor, have to be taken into consideration.

Recently van Oss and coworkers [2, 3] have introduced a new formulation of the surface and interfacial free energies. According to them, the energy is the sum of two components: apolar Lifshitz-van der Waals and polar acid-base of Lewis type. Then the work of adhesion of a liquid to a solid surface can be expressed as:

$$W_A = \gamma_L \left(1 + \cos\theta\right) = 2\left(\gamma_S^{LW}\gamma_L^{LW}\right)^{1/2} + 2\left(\gamma_S^+\gamma_L^-\right)^{1/2} + 2\left(\gamma_S^-\gamma_L^+\right)^{1/2} \tag{6}$$

where: γ^{LW} is the Lifshitz-van der Waals component, γ^+ and γ^- are electron-acceptor and electron-donor parameters of the acid-base component, respectively, and subscripts l and s stand for liquid and solid, respectively.

If one has measured contact angles of three liquids, whose surface tension components are known, then three equations of type (6) can be solved simultaneously and the surface free energy components of the solid can be determined. There is a hidden assumption that the strength of the interactions is the same irrespective of the probe liquid used, which is debatable.

McCarthy and coworkers [4], among others, stated that "single 'stationary' or advancing contact angle does not adequately describe the properties of a surface. Both the advancing and receding contact angles should be considered." The three-phase contact line is controlled by the topography or roughness of the surface, and hence by the contact angle hysteresis. In almost all practical solid-liquid systems, the contact angle hysteresis appears, even on the surface with the molecular-level roughness [5–7].

Lately Chibowski and coworkers [7–11] suggested a quantitative interpretation of the contact angle hysteresis (the difference between advancing θ_a and receding θ_r contact angles) for determination of surface free energy of a solid. The hysteresis is often explained as caused by roughness of real surfaces and/or the chemical micro-heterogeneities present on the solid surface [1]. Later it appeared that even on molecularly flat surfaces and on self-assembled monolayers the hysteresis emerges. Chibowski and coworkers [7–11] considered that contact angle hysteresis might also result from the liquid film present or left behind the droplet during retreat of its three-phase contact line. On the basis of this assumption, an equation to calculate the total surface free energy of a solid from three measurable parameters, i.e. advancing and receding contact angles and the liquid surface tension, was proposed.

Combining Equations (2–4) the liquid film pressure behind the drop reads:

$$\pi = \gamma_L \left(\cos \theta_r - \cos \theta_a \right) \tag{7}$$

The work of adhesion for both advancing and receding modes can be expressed as:

$$W_A^a = \gamma_L (1 + \cos \theta_a) \tag{8}$$

$$W_A^r = \gamma_L (1 + \cos \theta_r) \tag{9}$$

and hence the film pressure equals:

$$\pi = W_A^r - W_A^a \tag{10}$$

From Equation (10) it results that the film pressure can be expressed as the difference between the liquid work of adhesion to the film-covered surface and the work of adhesion of this liquid to the bare solid surface.

Based on Equations (7)–(10) an equation describing the total surface free energy of a solid results [7–10]:

$$\gamma_S^{tot} = \frac{\gamma_L (1 + \cos \theta_a)^2}{(2 + \cos \theta_r + \cos \theta_a)} \tag{11}$$

It should be noted that Equation (11) works also for zero receding contact angle, as well as when no hysteresis appears, i.e. $\theta_a = \theta_r$. In fact, even for a very low surface free energy solid, such as a nonpolar polymer, contact angle hysteresis occurs [8, 12–16], and this equation gives reasonable values of the surface free energy [8]. However, γ_S^{tot} values thus determined are apparent ones, because, to some extent, they depend on the kind and magnitude of the interactions taking place across the interface [7–10]. In other words, surface free energy thus determined depends on the kind of probe liquid used. It should be stressed that in this model assumption is made that the surface does not have to be molecularly smooth. It only assumes that the liquid film is left behind the drop during the retreat of the three-phase contact line. Also it is worth mentioning that lately Patankar [17] considered for rough surfaces "that the droplet leaves behind a thin film of liquid on the peaks of the pillars instead of leaving behind a dry surface". Also Neumann and coworkers [18] stated that "it is significant to note that hysteresis is not limited only to rough and heterogeneous surfaces. Theoretical modeling of contact angles on smooth and homogeneous surfaces also predicts contact angle hysteresis", and as one of the causes for the hysteresis they mentioned "penetration of liquid and swelling of the solid".

Extrand [15, 16] proposed a thermodynamic model for interpretation of contact angle hysteresis, which assumes that wetting can be considered as an adsorption-desorption process. Using contact angles data he calculated molar free energy ΔG of hysteresis from the equation:

$$\Delta G = -RT \ln (\theta_r/\theta_a) \tag{12}$$

where: R is the gas constant, T is the absolute temperature.

And the corresponding surface free energy (Δg) can be determined using the molar surface area A as:

$$\Delta g = \Delta G/A \tag{13}$$

The values thus calculated for nonpolar polymers, such as polyethylene (PE), poly(tetrafluoroethylene) (PTFE), polypropylene (PP), corresponded to the strength of dispersion (van der Waals) interactions, $\Delta G < 1$ kJ mol^{-1} [19]. The polar polymers gave higher free energies, $\Delta G \approx 1$–4 kJ mol^{-1}, which could be related to the internal energy resulting from conformational changes in the polar groups [15, 16].

It may be expected that on both low surface energy polymers and polar solid surfaces an adsorbed film of contact-angle liquids is present which influences the contact angles and the surface free energy. In the case of low surface energy polymers the film is probably not uniform and consists of molecular clusters and/or nano- and micro-droplets. Moreover, it may be expected that also the humidity of the environment affects the measured contact angles.

The purpose of this study was to determine the effect of increasing humidity of the ambient atmosphere on advancing and receding contact angles and thus on the surface free energy of PMMA as determined from the contact angle hysteresis. The molar free energies of the contact angle hysteresis were also calculated using a thermodynamic model proposed by Extrand [15, 16].

2. EXPERIMENTAL

2.1. Materials

The advancing and receding contact angles were measured on poly(methyl methacrylate) (PMMA) plates (5 cm^2), which were cut from a larger commercial plate (Dwory Oświęcim, Poland). The probe liquids used for contact angle measurements were water, formamide (Fluka, >99%) and diiodomethane p.a. (POCH S.A., Poland). The water was from Milli-Q Plus 185 and then was distilled using Destamat Bi18E system (Heraeus Quarzglas, Germany) (pH = 7.25).

2.2. Sample preparation

Before the measurements the PMMA plates, after removing the protective foil, were washed in an ultrasonic bath for 15 min, first with 20% methanol p.a. grade (POCH S.A., Poland) and then three times with water from Milli-Q 185 system. The plates after washing were dried at 60°C for 2 h and kept in a desiccator.

2.3. Contact angle measurements

Digidrop GBX Contact Angle Meter (France) with video-camera system and computer software was used for the contact angle measurements by the sessile drop method. The advancing contact angle of the liquid was measured using 6 μl droplet, which was gently deposited on the surface. Then after sucking of 2 μl from the droplet into the syringe the receding contact angle was measured.

The contact angles of water were measured at room temperature (20 ± 1°C) in a closed chamber at relative humidity ranging from 1.5 to 100%, taking readings on the left and right sides of 20 droplets. At selected humidities (1.5, 50 and 100%) contact angles of diiodomethane and formamide were also determined. The humidity was controlled by the humidity generator installed in the Contact Angle Meter and maintained constant within ±0.1%.

Before contact angle measurements, the PMMA plates were placed in the measuring cell of the apparatus with controlled humidity and were conditioned for 5 min in a given humidity. For the surface free energy calculations the literature values for surface tension and its components of the probe liquids were taken [3].

2.4. Atomic force microscopy (AFM)

The PMMA surface was investigated using atomic force microscopy (AFM, Nanoscope 3, VEECO).

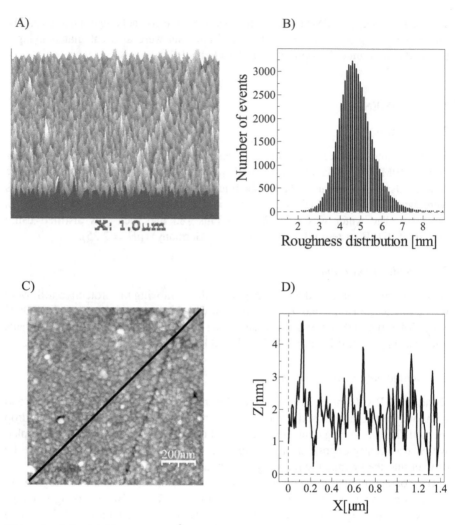

Figure 1. A) 3D AFM image of 1 μm^2 PMMA surface, B) roughness distribution of the surface, C) 2D AFM image of the same surface, D) roughness along the line shown in C) from top to bottom.

3. RESULTS AND DISCUSSION

To learn about the heights of protrusions and their distribution on the surface of PMMA sample the AFM images were taken in several places of the surface in the contact mode with a silicon tip. The force on the tip was repulsive with a mean value of 10^{-9} N. Figure 1 shows 3D and 2D images (Figs. 1A and 1C, respectively) of the 1 μm^2 surface, and the roughness distribution of this surface (Fig. 1B), as well as the roughness heights (Fig. 1D) along the marked line in Fig. 1C. From these

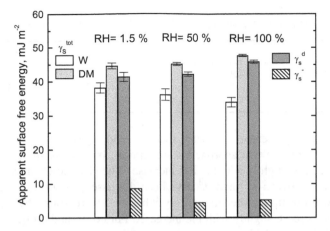

Figure 2. Apparent total surface free energy of PMMA determined from the contact angle hysteresis of diiodomethane (DM) and water (W), and apolar γ_S^d and polar electron-donor γ_S^- components of the surface free energy at various relative humidities.

images it is found that on 94.9% of the surface the protrusion heights are between 2.12 and 5.70 nm, and only 3.2% of this surface possesses 5.83–8.75 nm high protrusions. The average protrusion of the roughness is 4.9 nm and the root mean square (RMS) roughness is 0.9 nm. These data were obtained using WSxM 4.0 Develop 8.0 Scanning Probe Microscope software [20], and they indicate that the PMMA surface is quite smooth, although the protrusions are rather sharp (Fig. 1A).

The contact angle data on PMMA surface and its surface free energy determined from van Oss and coworkers [2, 3] (Lifshitz/van der Waals acid–base components, LWAB) and contact angle hysteresis (CAH) approaches [7–10] can be found in the literature [7, 9, 21–26]. Therefore, the values of the components and total surface free energy for PMMA obtained in this paper could be compared with results from the literature.

Figure 2 shows the total surface free energy of PMMA at three relative humidities of the atmosphere, i.e. 1.5, 50 and 100%, as determined from the contact angles of diiodomethane and water using CAH [7–10] approach. In the figure are also shown the values of apolar dispersion component γ_S^d, which is the same as γ_S^{LW} in the LWAB model [2, 3], and the electron-donor parameter γ_S^- of the surface free energy calculated from the advancing contact angles of diiodomethane, water and formamide from the LWAB approach. For the calculations the literature data for the surface tension and its components of the probe liquids were used and they are shown in Table 1 [2, 3].

Generally the values in Fig. 2 are in agreement with the literature ones [7, 9, 22–26]. PMMA surface is a monopolar one, i.e. it shows only the electron-donor interactions, but no electron-acceptor parameter is detected [7, 9, 22–26].

Table 1.
Surface tension components of probe liquids (mN/m) used for contact angle measurements

Liquid	γ_L	γ_L^{LW}	γ_L^-	γ_L^+
Diiodomethane (DM)	50.8	50.8	–	≈ 0
Water (W)	72.8	21.8	25.5	25.5
Formamide (F)	58.0	39.0	39.6	2.28

From the results presented it appears that γ_S^{tot} and γ_S^d of PMMA obtained from diiodomethane by the two approaches are similar, but γ_S^{tot} is slightly larger than γ_S^d. Both increase slightly with the RH increase. The physico-chemical reasons for the difference between γ_S^{tot} and γ_S^d have been discussed elsewhere [7, 27] and the algebraic relationship between them is given below. The reasons why the dispersion interactions increase slightly with the increased humidity (and probably increasing amounts of the adsorbed water vapor) are not known yet. This will need more detailed studies and some causes for it can be found in the literature and will be discussed later [27–30]. The difference between γ_S^{tot} and γ_S^d determined with diiodomethane contact angles may result from different interacting distances of the liquid/solid molecules [27]. It should be noted that the differences are in all cases larger than the standard deviations shown in Fig. 2.

It was concluded [10] that the surface free energy of a solid surface if determined via contact angles of a probe liquid was an apparent quantity, which varied depending on the kind of liquid used. Using the Fowkes approach [28] and Young-Dupre equation (5), and if only the dispersion interaction with a liquid takes place across the interface, for which $\gamma_L = \gamma_L^d$ (n-alkane and diiodomethane), γ_S^d can be calculated as [10, 27]:

$$\gamma_S^d = \gamma_L \left(1 + \cos\theta_a\right)^2 / 4 \tag{14}$$

Then the relationship between γ_S^{tot} and γ_S^d can be derived from Equations (11) and (14) as:

$$\gamma_S^{tot} = \gamma_S^d \left(1 + \frac{2 - \cos\theta_a - \cos\theta_r}{2 + \cos\theta_a + \cos\theta_r}\right) \tag{15}$$

From Equation (15) it is seen that both advancing and receding contact angles determine the difference between γ_S^{tot} and γ_S^d.

Comparing the total surface free energy γ_S^{tot} of PMMA determined from the CAH approach and nonpolar diiodomethane and polar water contact angles (Fig. 2), it is seen that the γ_S^{tot} values determined at various RH values with diiodomethane are larger by a few mJ m^{-2} than γ_S^{tot} determined with water. These results show, as was mentioned above, that the solid surface free energy depends on the probe liquid surface tension, i.e. the kind and strength of liquid-solid interfacial interactions, and, therefore, such values should be treated as apparent ones.

From the advancing contact angles of diiodomethane and water on PMMA surface also the work of adhesion was calculated from Equation (5) and the values are listed

Table 2.
Values of advancing and receding contact angles of diiodomethane (DM) and water (W) on PMMA surface at three relative humidities, work of adhesion and molar free energy of hysteresis of these liquids

Relative humidity	θ_a/θ_r, deg		W_A, mJ m^{-2}		ΔG, mJ mol^{-1}	
	DM	W	DM	W	DM	W
1.5%	36.2/24.7	81.1/67.4	91.8	84.1	0.931	0.450
50%	34.5/23.5	83.2/66.3	92.7	81.4	0.935	0.491
100%	25.8/19.2	86.0/67.5	96.5	77.9	0.720	0.605

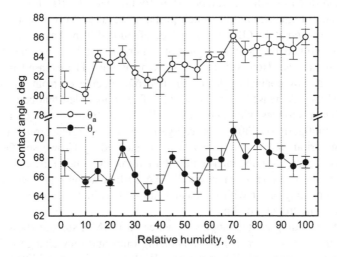

Figure 3. Advancing and receding contact angles of water droplets on PMMA vs. relative humidity of the ambient atmosphere.

in Table 2, together with molar free energies of hysteresis calculated from Equation (12) [15, 16].

Irrespective of relative humidity the work of adhesion of diiodomethane is greater than that of water. The same is true for molar free energies of hysteresis for these liquids [15, 16]. For diiodomethane ΔG varies from 0.931 to 0.720 kJ mol^{-1} and for water from 0.450 to 0.605 kJ mol^{-1}, i.e. for nonpolar diiodomethane molar free energies of hysteresis are larger than for polar water [15]. Across the diiodomethane/PMMA interface the interactions are solely of dispersion nature and they are stronger than the dispersion and acid-base interactions across water/PMMA interface. It should be remarked that γ_L^d equals 50.8 mJ m^{-2} for diiodomethane and is only 21.8 mJ m^{-2} for water, and the polar interactions of PMMA are relatively weak, i.e. 5–8 mJ m^{-2} (Fig. 2).

To find the effect of relative humidity on water contact angles and, consequently, on PMMA surface free energy the advancing and receding contact angles of water were measured at various relative humidities, from 1.5 to 100% RH, and they are shown in Fig. 3. As can be seen if the humidity increases from 1.5 to 100% the advancing contact angles also increase from 80° to 86°, but they fluctuate depending on the RH. The absolute errors calculated for 95% confidence level were below 2 degrees, which is typical for contact angle measurements. Similar relationship of advancing contact angles of water with increasing RH was observed by Mackel and coworkers [29] on thin films ($\approx$ 1 μm) of poly(ethylene glycol) (PEG) with fluoroalkyl endgroups (6 kg/mol PEG with 10-carbon fluoroalkyl, denoted as 6KC10) deposited on silicon wafers. They found that 6KC10 films became more hydrophobic at higher humidity, which probably resulted from a reorganization of the groups exposed at the 6KC10 gel surface of studied film. Yasuda and coworkers [30] also examined the contact angles of water on a gelatin gel and agar gel at different humidities. However, in this system, as the RH increased from 28% to 90% the advancing contact angles on the hydrogel decreased from 90° to 75°, and for agar gel they decreased from 34° to 22°. They concluded that better wetting of the biopolymer surfaces with increasing RH resulted from a change in the surface molecular configuration [30]. The reorganization of surface groups on polymer surfaces in various solvents has also been considered by Crowe and Genzer [31], which caused changes in the interfacial interactions and thus in the contact angle of water. One can also consider that on PMMA surface some rearrangements of the surface groups may occur with increasing humidity which can interact with water molecules via hydrogen bonding.

As for the receding contact angles of water on PMMA surface shown in Fig. 3, first of all they are smaller than the advancing ones and they change as a function of RH run in a somewhat different way. With increasing humidity a significant minimum (at 35% RH) and maximum (at 70%) are seen. The observed changes in the advancing and receding contact angles are evidently due to the water vapor adsorption on the PMMA surface. The structure of the water film depends on the adsorbed amount, and it first consists of the water clusters, and then of nano- or micro-droplets.

The resulting contact angle hysteresis versus relative humidity is presented in Fig. 4. Three distinct regions can be observed. The lowest contact angle hysteresis appears in the RH range from 1.5 to 10% but it increases roughly linearly up to RH $\approx$ 20%. Then in the range between 20% and 70% RH the values of hysteresis oscillate quite periodically between 15.2° and 17.2°. Finally, for relative humidity from 70% to 100% the hysteresis of water contact angle increases from 15.2° to 18.5°.

Erbil and coworkers [32] using video-microscopy determined the advancing and receding contact angles on PMMA and poly(ethylene terephthalate) (PET) by evaporation of the sessile drop of water and measurements of the contact diameter of the deposited droplet. The measurements were performed in a closed chamber at

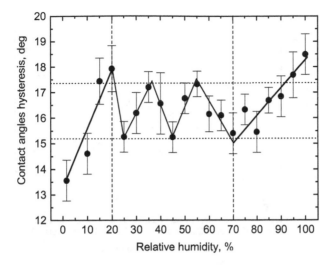

Figure 4. Contact angles hysteresis of water droplets on PMMA vs. relative humidity of the ambient atmosphere.

constant temperature (22.8±0.2°C) and relative humidity (41.8±0.8%) inside the chamber. They found the average hysteresis on PMMA surface to be 23.5±1.5°, which is a few degrees larger than in our study (ca. 16.5°) (Fig. 3). One of the reasons for the difference may be the different origins of the PMMA samples and the differences in the surfaces roughness and topography. This is also reflected in the values of advancing contact angles of water measured on a PMMA surface (Fig. 3), which are higher than the literature values ranging from 62.2° to 80° [7, 9, 21–26, 34, 35].

It is worth mentioning that in the calculations of the solid surface free energy from advancing contact angles the presence of any film behind the drop is neglected [1, 21, 34]. However, it is reported that even on low surface energy polymers the liquid film is present and may play a significant role in the interfacial interactions [27, 36]. Busscher and coworkers [37, 38] affirmed that even if $\gamma_L > \gamma_S$ the spreading pressure can have considerable influence on the contact angle. Based on the correlation between water or propanol adsorption on various surfaces and spreading pressures, they found that the film pressures were of the same order of magnitude for these liquids on both high- and low-energy surfaces, despite the surface tension values of propanol and water being quite different (23.7 and 72.8 mN m^{-1}, respectively).

On the other hand, van Oss and coworkers [25] evaluated the possible maximal effect of the liquid vapor adsorption behind the drop on the contact angles on PMMA for commonly applied probe liquids (water, 1-bromonaphthalene, diiodomethane, formamide, ethylene glycol, glycerol), and for water on poly(ethylene oxide) (PEO).

L. Holysz et al.

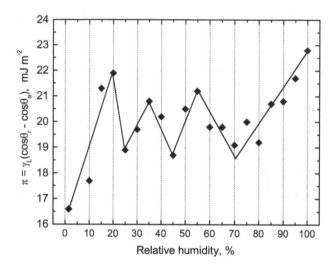

Figure 5. Water film pressure, π on PMMA calculated from Equation (7) vs. relative humidity of the ambient atmosphere.

Using the data of saturated vapor pressures of these liquids at 20°C and 760 mm of Hg atmospheric pressure, the work of adhesion and cohesion, and applying the Langmuir adsorption equation and Cassie's approach, they found that only water and ethylene glycol significantly affected the contact angles on PMMA. Thus, the measured contact angles would be lower than those on the bare surface, as a consequence of the adsorbed molecules originating from the liquid droplets (only by about 1.55° and 1.07°, for water and ethylene glycol, respectively).

But Fowkes [28] showed that a single physically adsorbed water monolayer on quartz already reduced its dispersion component of the surface free energy from 76 to 29.7 mJ m^{-2}. Three adsorbed water monolayers already caused the dispersion component to be almost equal to that of bulk water (21.8 mJ m^{-2}) [28, 39]. However, up to 10 statistical water monolayers could be physically adsorbed onto quartz surface. This is because the surface free energy of the outermost adsorbed layer is still slightly less than that of bulk water. Obviously, the equilibrium film thickness depends on the kind and strength of solid/liquid interactions.

Applying the CAH model to our results (Equation (7)) the water film pressure on PMMA surface was calculated depending on the RH and the values are shown in Fig. 5. These changes resemble those of contact angle hysteresis (Fig. 4). The water film pressure changes periodically with increasing relative humidity. The largest changes and roughly a linear increase of the film pressure occurs first between 1.5% and 20% RH, from 16.5 to 22 mJ m^{-2}. Then, clearly visible minima appear at 25, 45 and 70% RH, while at 20, 35 and 55% RH maxima are seen. At RH > 70% again a linear increase in π is observed between 70% and 100% of RH. The highest value

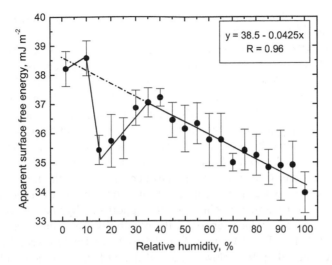

Figure 6. Changes in apparent surface free energy of PMMA depending on relative humidity of the ambient atmosphere.

of the film pressure (ca. 23 mJ m^{-2}) occurs at 100% RH (Fig. 5). Total increase in the film pressure of water is from 16.5 mJ m^{-2} (at 1.5% RH) to 23 mJ m^{-2} at 100% RH.

As discussed above the apparent total surface free energy of PMMA can be evaluated via contact angle hysteresis approach from Equation (11). The calculated apparent values of γ_S^{tot} as a function of relative humidity are shown in Fig. 6. It appeared that 1.5–10% RH practically does not affect γ_S^{tot}. Then it decreases sharply at 15% RH from 38.5 to 35.5 mJ m^{-2}, to linearly increase to ca. 37.5 mJ m^{-2} at 35% RH. For greater RH values up to saturated vapor pressure, the surface free energy decreases linearly ($R = 0.96$). From the linear fit (Fig. 6) extrapolation to RH = 0% the surface free energy of PMMA equals ca. 38.5 mJ m^{-2}, which is actually equal to the average γ_S^{tot} obtained at 1.5–10% RH.

At the present stage of the study, the interpretation of the obtained apparent surface free energy changes of PMMA caused by increasing humidity is difficult. These values should be recognized as those reflecting the surface free energy of bare PMMA surface on which some amount of water is adsorbed from the ambient atmosphere. The linear decrease of the surface free energy between 35–100% RH can be interpreted as caused by the decreasing electron-donor γ_S^- interactions, which was found from the LWAB approach (Fig. 2). Taking for water $\gamma_L^- = 25.5$ mJ m^{-2} [2, 3] and assuming a decrease in γ_S^- of PMMA of 0.1 mJ m^{-2}, the resulting decrease, $\Delta\gamma_S$, in the apparent surface free energy would amount to $2(25.5 \times 0.1)^{1/2} = 3.2$ mJ m^{-2}, which is just equal to the γ_S^{tot} decrease in the discussed RH range, i.e. 35–100% (Fig. 6). It seems that most intriguing are the apparent γ_S^{tot} changes in the 1.5–35% RH range. At 15% RH the γ_S^{tot} value is comparable to that at ca. 75% RH.

At very low RH values (1.5 and 10%) the effect of water vapor on the γ_S^{tot} cannot be explained, because the density of the adsorbed water clusters (or nano-droplets) is rather very low. The sharp decrease in γ_S^{tot} by 3.5 mJ m^{-2} at RH between 10 and 15% indicates that the adsorbed water affects the three-phase contact line and thus the contact angle. In other words, the film is "seen" by the deposited droplet of water. Following this idea, with increasing RH more and more adsorbed water on the surface should be present. As water does not spread over the surface, so depending on the adsorption density it forms nano-clusters and eventually micro-droplets. The water droplet (6 μl volume) deposited from the microsyringe may capture the adsorbed water molecules, not only those present beneath it, but also possibly those some distance behind it. To support this hypothesis we have done a 'macro experiment', which supported the possibility that such mechanism might occur indeed. On the PMMA surface four 0.3 μl water droplets were deposited and then a large water drop 9 μl slowly touched the surface with the small deposited droplets. The whole process was filmed. After contacting the large water drop with small ones, it absorbed all the four smaller droplets. But, more important was the clearly visible retreat of the edges of the drop formed, whose volume had increased from 9 μl to 10.2 μl. Three characteristic frames of the film, taken after 0.0, 2.22 s, and 3.0 s, are shown in Fig. 7, which illustrate the process described above. This experiment showed that "self-cleaning" effect on the PMMA surface takes place. As the density of the water nano- or micro- droplets on the surface will vary depending on the RH value, they can coalesce with each other and similar situation on the surface can occur periodically with increasing RH of the ambient

A)

B)

C)

Figure 7. Selected frames from the film showing deposition of large water drop (9 μl) on PMMA surface on which four smaller water droplets (0.3 μl) had been first deposited: A) first frame (0 s), B) sixty eighth frame (2.22 s), C) eighty first frame (3 s).

atmosphere. Of course, this hypothesis needs further study, but such a mechanism could explain the observed non-linear, periodical, changes in the advancing and receding contact angles, as well the resulting contact angle hysteresis and the film pressure (Figs. 3–5).

Finally, from the changes in the surface free energy of PMMA on which the adsorbed water film (clusters, nano- or micro-droplets) is present, γ_{Sf} can be calculated. There should be a distinct difference between the apparent γ_S^{tot} values calculated from Equation (11) and the γ_{Sf} values calculated from Equation (4). The γ_{Sf} values can be considered as those resulting when "an added-film of water" is present (over that formed by water vapor adsorption from the ambient atmosphere). These γ_{Sf} surface free energy values are described by the receding contact angles (Equation (3)). So, the γ_{Sf} surface free energy of PMMA can be obtained by adding the respective π values, i.e. by adding the results from Fig. 5 to those of Fig. 6. The values of the γ_{Sf} surface free energy of PMMA thus obtained are plotted in Fig. 8 versus relative humidity values. The changes in γ_{Sf} are quite different from those in γ_S^{tot} shown in Fig. 6. They are rather a consequence of the film pressure changes shown in Fig. 5. At RH = 1.5% the apparent $\gamma_S^{tot} = 38$ mJ m^{-2} and the $\gamma_{Sf} = 55$ mJ m^{-2}. The difference is due to the water film formed behind the drop after the three-phase line has retreated. The γ_{Sf} values oscillate between 54 and 58 mJ m^{-2} and they can be considered as a result of overlapping of two effects: i) the increasing adsorption density of water and changes in film structure (cluster, nano-, micro-droplets, and the 'self-cleaning' mechanism), and ii) the water film left behind the drop upon its retreat. However, as the γ_{Sf} correlates with the film pressure π

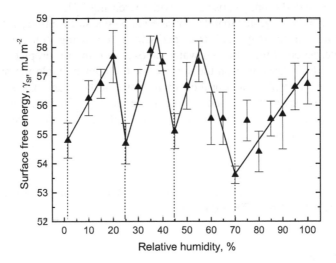

Figure 8. Changes in the surface free energy of PMMA with water film depending on the relative humidity of the ambient environment.

110   *L. Holysz* et al.

(Figs. 8 and 5) it may be concluded that π values determine the PMMA surface free energy γ_{Sf} on which the water film is left behind the drop. On the other hand, the film pressure is determined by the difference in cosines of advancing and receding contact angles (Equation (7)), whose values result from the total energy balance in the solid (film)/water droplet/air system. The energy balance varies depending on the state of the water film adsorbed.

The analogous experiments with silicon, which are under way, show similar relationships as those described above. They should shed more light on these intriguing results.

4. CONCLUSION

Even on molecularly flat PMMA surface (RMS roughness = 0.9 nm) the contact angle hysteresis appears. Both, the advancing and receding contact angles vary depending on the relative humidity ranging from 1.5% to 100%, but the changes are not monotonic. Hence the contact angle hysteresis also changes periodically and generally increases from 13.5° (RH = 1.5%) to 18.5° at RH = 100%. The greatest increase in the hysteresis takes place at RH = 1.5–20% and then at RH = 70–100%.

The hysteresis changes are reflected in the free energy changes. The PMMA apparent surface free energy γ_S^{tot} in the RH range 35–100% determined from the CAH model using water contact angles decreases with increasing humidity. However, at lower RH values it shows a striking minimum at 15% RH. On the other hand, the calculated surface free energy of PMMA surface with water film left behind the retreated drop, γ_{Sf}, oscillates with increasing RH.

A mechanism is suggested that the observed apparent surface free energy changes of the PMMA surface may result from periodically occurring 'self-cleaning' effect due to coalescence of the adsorbed water nano- and/or micro- droplets. During the larger water drop deposition on the surface for the contact angle measurements it absorbs the nanodroplets, and at their high densities they can coalesce with each other. As a consequence, the solid surface in the vicinity of the there-phase contact line becomes free of the nano- and/or micro-droplets. However, this hypothesis needs further verification.

Acknowledgements

We are grateful for financial support from the Ministry of Science and Higher Education (Project No. 3 T09A 043 29).

REFERENCES

1. A. W. Adamson and A. P. Gast, *Physical Chemistry of Surfaces,* Sixth Edition, Wiley, New York (1997).
2. C. J. van Oss, M. K. Chaudhury and R. J. Good, *J. Chem. Rev.* **88**, 927 (1988).

3. C. J. van Oss, *Colloids Surfaces A* **78**, 1 (1993).
4. W. Chen, A. Y. Fadeev, M. C. Hsieh, D. Öener, J. Youngblood and T. J. McCarthy, *Langmuir* **15**, 3395 (1999).
5. A. Y. Fadeev and T. J. McCarthy, *Langmuir* **15**, 3759 (1999).
6. A. Y. Fadeev and T. J. McCarthy, *Langmuir* **16**, 7268 (2000).
7. E. Chibowski, A. Oniveros-Ortega and R. Parea-Carpio, *J. Adhesion Sci. Technol.* **16**, 1367 (2002).
8. E. Chibowski, in: *Contact Angle, Wettability and Adhesion*, Vol. 2, K. L. Mittal (Ed.), pp. 265–288. VSP, Utrecht (2002).
9. H. Radelczuk, L. Hołysz and E. Chibowski, *J. Adhesion Sci. Technol.* **16**, 1547 (2002).
10. E. Chibowski, *Adv. Colloid Interface Sci.* **103**, 149 (2003).
11. E. Chibowski, L. Hołysz, A. Zdziennicka and F. González-Caballero, in: *Surfactants in Solution*, A. K. Chattopadhyay and K. L. Mittal (Eds), pp. 31–53. Marcel Dekker, New York (1996).
12. C. Della Volpe, A. Deimichel and T. Ricco, *J. Adhesion Sci. Technol.* **12**, 1141 (1998).
13. J. Höpken and M. Möller, *Macromolecules* **25**, 1461 (1992).
14. L. M. Lander, L. M. Siewierski, W. J. Brittain and E. A. Volger, *Langmuir* **9**, 2237 (1993).
15. C. W. Extrand, *J. Colloid Interface Sci.* **202**, 462 (1998).
16. C. W. Extrand, in: *Contact Angle, Wettability and Adhesion*, Vol. 2, K. L. Mittal (Ed.), pp. 289–297, VSP, Utrecht (2002).
17. N. A. Patankar, *Langmuir* **19**, 1249 (2003).
18. H. Tavana, D. Jehnichen, K. Grundke, M. L. Hair and A. W. Neumann, *Adv. Colloid Interface Sci.* (2007) doi: 10.1016/j.cis.2007.04.008.
19. J. N. Israelachvili, *Intermolecular and Surfaces Forces*, 2nd Edition, Accademic Press, New York (1992).
20. http://www.nanotec.es.
21. C. J. van Oss, R. J. Good and H. J. Busscher, *J. Dispersion Sci. Technol.* **11**, 75 (1990).
22. B. Jańczuk, W. Wójcik and A. Zdziennicka, *J. Colloid Interface Sci.* **157**, 393 (1993).
23. M. Jurak and E. Chibowski, *Langmuir* **22**, 7226 (2006).
24. E. Chibowski, L. Hołysz, K. Terpiłowski and M. Jurak, *Colloids Surfaces A* **291**, 181 (2006).
25. C. J. van Oss, R. F. Giese and W. Wu, *J. Dispersion Sci. Technol.* **19**, 1221 (1998).
26. L.-H. Lee, *Langmuir* **12**, 1681 (1997).
27. E. Chibowski, *Adv. Colloid Interface Sci.* **113**, 121 (2005).
28. F. M. Fowkes, in *Hydrophobic Surfaces*, F. M. Fowkes (Ed.), p. 151. Academic Press, New York (1969).
29. M. M. Mackel, S. Sanchez and J. A. Kornfield, *Langmuir* **23**, 3 (2007).
30. T. Yasuda, T. Okuno and H. Yasuda, *Langmuir* **10**, 2435 (1994).
31. J. A. Crowe and J. Genzer, *J. Am. Chem. Soc.* **127**, 17610 (2005).
32. H. Y. Erbil, G. McHale, S. M. Rowan and M. I. Newton, *Langmuir* **15**, 7378 (1999).
33. M. L. Tate, Y. K. Kamath, S. P. Wesson and S. B. Ruetsch, *J. Colloid Interface Sci.* **177**, 579 (1996).
34. C. Della Volpe, D. Maniglio and S. Siboni, in: *Contact Angle, Wettability and Adhesion*, Vol. 2, K. L. Mittal (Ed.), pp. 45–71. VSP, Utrecht (2002).
35. R. Bongiovanni, V. Lombardi and A. Priola, in: *Contact Angle, Wettability and Adhesion*, Vol. 2, K. L. Mittal (Ed.), pp. 101–108. VSP, Utrecht (2002).
36. B. C. Nayar and A. W. Adamson, in: *Physicochemical Aspects of Polymer Surfaces*, Vol. 2, K. L. Mittal (Ed.), pp. 613–623, Plenum Press, New York (1983).
37. H. J. Busscher, A. W. J. van Pelt, H. P. de Jong and J. Arends, *J. Colloid Interface Sci.* **95**, 23 (1983).
38. H. J. Busscher, A. G. M. Kip, A. van Silfhout and J. Arends, *J. Colloid Interface Sci.* **114**, 307 (1986).
39. P. Staszczuk, B. Jańczuk and E. Chibowski, *Mater. Chem. Phys.* **12**, 469 (1985).

Part 2

Relevance of Wetting
in Cleaning and Adhesion

Contact Angle, Wettability and Adhesion, Vol. 5, pp. 115–138
Ed. K.L. Mittal
© VSP 2008

Wettability measurements and cleanliness evaluation without substantial cost

JOHN B. DURKEE[1,*] and ANSELM KUHN[2]

[1] *P.O. Box 847, Hunt, TX 78024, USA*
[2] *105 Whitney Drive, Stevenage, Herts SG1 4BL, UK*

Abstract—Cleanliness can be characterized in industrial applications via "simple" wettability measurements, and has been successfully done so for at least two centuries. A problem in much of general manufacturing and maintenance industries is not that more sophisticated measurement and evaluation technology is necessary to provide value, but is that technology developed at least several generations ago has not been more widely and profitably used. This paper describes and references that technology, and identifies published case histories where it has been both successfully and unsuccessfully used.

Keywords: Surface cleanliness; wettability; contact angle; surface energy; Nordtest method; Bikerman equation; low cost.

1. INTRODUCTION

The characterization of surface cleanliness can be done using a wide variety of techniques. Each should be chosen with recognition of the characteristics of the native surface and the soil being removed by the cleaning process [1].

Surface cleanliness is not an absolute benchmark. It is simply a condition which allows the object with that surface characteristic to be used as desired. Consequently, in this paper, cleanliness is a relative measurement appropriate to the application. In other words, the focus of this paper is not absolute measurements of surface character. It is about measurements or observations of surface character relative to some previously established standard by which cleanliness is defined [1].

*To whom correspondence should be addressed. Tel.: (830) 238-7610; Fax: (612) 677-31170;
e-mail: jdurkee@precisioncleaning.com

2. TWO APPROACHES TO CLEANLINESS EVALUATION

Often satisfactory cleanliness of a surface is inferred when the next use to which that surface is put is known to be satisfactory. While that may (or may not) be an acceptable industrial practice, it is not a measurement of surface cleanliness, as:

- Some tests determine a total level and type of contamination without identifying it by chemical content or location. Examples are non-volatile residue (NVR) and total organic carbon (TOC). The chemical content is identified in other surface tests. Still other cleanliness tests are performance standards such as peel strength or surface corrosion.

- Other tests determine a local (topical) level of cleanliness without identifying it by chemical content. They evaluate the wettability of a liquid at a specific location on a surface. Sampling is a critical issue because wettability at any test point may not represent the average condition of the surface. The chemical content is identified in other surface tests.

It is the facilities and techniques useful in measuring wettability which are the subject of this paper. Of specific interest are those techniques and facilities which allow management of surface cleaning with limited infrastructure and at low cost.

3. WETTABILITY AND SURFACE CLEANLINESS

A soiled surface may be characterized by the behavior of a drop of liquid when applied to it. Said another way, the cleanliness of a surface may be inferred by the conformation (shape) of a drop of liquid applied to the surface.

The drop shape (contact angle between the drop and the surface) will be determined by:

A. The characteristic of the liquid applied to the surface, which is likely known.

B. The energetic characteristic of the solid surface, i.e., what is wanted to be known (*because it is being taken as a proxy for cleanliness*) [2].

C. The characteristic of the solid surface as it interacts with a vapor phase (usually air); this characteristic may be uncertain.

3.1. Evolution of the relationship between wettability and cleanliness

Within forty years of Young's publication on wetting phenomenon, scientists in industry had harnessed these ideas in order to optimize a range of processes [3].

A. In 1843, Ludwig Moser [4, 5] published two papers purporting to show how clean glass could be differentiated from soiled glass – depending on how moisture condensed on it. This is one of the first embodiments of the water-break test.

B. In 1902, Pockels [6] proposed that a cleanliness test could be established based on how talc interacted with a soiled surface.

C. In 1923, Devaux [7] expressed a preference for what is still considered a valued and commonly used technique: the water-break test.

D. Technology in use today was foreshadowed by Baker and Schneidewind [8] in 1924 when they reported that cleaning efficiency was evaluated by "measuring relative interfacial tensions."

E. Nelson [9] expressed the same point of view in 1943 after cleaning lubricating oil from steel.

Perhaps the key point of this paper should be that, like politics, all cleanliness is local. No single technique for measuring cleanliness is always superior. It is the characteristics of the soil and the parts, and the details of next use which determine all [10].

Zakarias [11] validated this view in 1936, when he preferred what was apparently the "white glove test" to a specific analytical evaluation of surface residue and some variant of the water-break test.

Wettability [12] represents a practical approach to measuring surface cleanliness. Instead of visual examination of the native surface, using analytical tests to identify the level and location of chemical contamination, or conducting a standard test such as peel strength, which may be a proxy for cleanliness, some users evaluate the cleanliness of a surface by placing a tiny amount of liquid on it and watching how it wets the surface.

The reasons wettability measurements are practical, low-cost, and easy to implement, and *do* define cleanliness, are:

A. Testing is non-destructive.

B. Investment in facilities is minimal.

C. Evaluation can be done within minutes.

D. Numerous case studies, discussed below, can serve as guides.

4. CLEANLINESS EVALUATION WITH LIMITED EQUIPMENT

Ideas developed nearly two centuries ago can provide improved accuracy and repeatability, and automate the examination so more surface area can be evaluated. Useful, credible, and valuable evaluations of surface cleanliness can be made without any equipment, and separately with equipment probably recognized by Ludwig Moser (see below). These will be described in the remainder of this paper. Published industrial examples in which they were successfully employed will be related.

There are *two* significant uses of cleanliness tests. The obvious one is to learn whether the parts are clean enough to be passed on to subsequent process steps. The other is to learn whether too much time, space, and money have been spent on cleaning work when a lesser amount of resources would have allowed subsequent processing to be satisfactorily done.

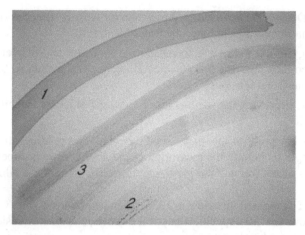

Figure 1. Outcomes with use of dyne liquids. Courtesy of Sherman treaters.
(http://www.shermantreaters.co.uk)

4.1. Surface tension test liquids

Platers, printers [13, 14], coaters, glaziers, and other operators are very concerned
about coverage of liquids on solid surfaces. After all, if ink, paint, or a finish does
not wet a surface, it cannot adhere to it via adhesion or cover it via adsorption.

That is why surface tension test liquids, also known as dyne liquids or test inks,
were developed. They are most frequently used to bracket the surface energy level
of polymer or paper substrates. Surface tension test liquids are binary prepared
(formulated) mixtures whose surface tension is known. They are portable surface
tension standards. They can be used to establish surface characteristic – cleanliness
relative to a standard [15].

Dyne liquids are applied [16] as a continuous film to about 1 square inch (6.5 cm^2)
of the surface under study by a brush, swab, wick, or felt-tipped pen. Care must be
taken in their application so that mechanical cleaning work is not done by moving or
removing soil by contact with the applicator. The operator carefully observes when
and if the continuous film *retracts* and breaks up into droplets. So, the observation
is of the liquid conformation and not of a contact angle between the film and the
surface.

The applied continuous film can experience three possible outcomes, which are
demonstrated in Fig. 1 [2].

1. The film *remains* continuous for at least 2 seconds (label 1 in Fig. 1). Experience
   has shown that wetting is normally adequate when the continuous film of dyne
   liquid *remains intact* for at least 2 seconds. This means that the surface tension
   of the test ink is *less* than the surface energy of the surface.
2. The film partially or incompletely decays into droplets. Both partial and
   incomplete outcomes are shown as label 3 in Fig. 1. Some continuous film

remains, with perhaps droplet formation at the film edge. This means that the surface tension of the test ink is *barely above* the surface energy of the surface.

3. The film *quickly* decays into droplets – almost immediately as applied (label 2 in Fig. 1). This means that the surface tension of the test ink is *greater or considerably greater* than the surface energy of the surface.

In other words, a liquid cannot wet or continuously coat a surface unless its surface energy is *less than* that of the surface. Wetting can be thought of as the process of achieving molecular contact.

Normally users *start* with a test liquid believed to have a *higher* surface tension than the surface energy of the surface. Here they see droplets, rivulets, or globules. They do not see a continuous film [17]. Then test liquids with incrementally *lower* surface tensions are tried on different areas. The lower the surface tension of the test liquid the higher its wetting ability.

The desired outcome is when a test liquid does form a continuous film but does not form a discontinuous film or discrete droplets, rivulets, or globules. It is conventional to say the surface has a "dyne value" equal to the surface tension of the solution which maintained a continuous film.

In this way, the surface energy of the test surface is estimated, or bracketed, without equipment. Here surface energy is taken as the proxy for cleanliness. Surface energy of metal surfaces is usually increased by light abrasion or treatment with an acid etching solution so that metal oxides are removed.

Said another way, the level of cleanliness of the test surface has been estimated or bracketed. Comparison should then be made with the level of surface energy measured using the surface of a piece (part) known to perform properly in downstream operations.

At least four international standards cover this determination, chiefly for printing applications on fibers, papers, or films. They are ASTM D2578-04a [18], ISO 8296 [19], JIS K6768 [20], and DIN 53 364 [21].

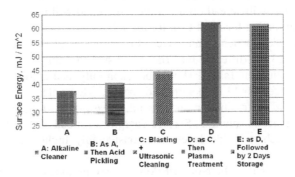

Figure 2. Surface energy values as determined by dyne liquids.

4.1.1. Case history with use of dyne liquids to measure cleanliness

An excellent example is the work of Stiles *et al.* [22]. Their results, which were obtained using various methods of cleaning rough steel surfaces, are plotted in Fig. 2. These results are plausible in that plasma treatment provides a substantially clean surface, and that acid pickling provides only a modest advantage over alkaline aqueous cleaning.

This work shows how dyne liquids, a low-cost technology, can be used to assess the effects of various methods of surface cleaning.

4.2. The Nordtest Poly 176 method

This is a quantitative method developed for cleaning evaluations. Its proper use includes a control state.

The Nordtest Poly 176 [23] method uses liquids which are normally prepared by the user. They are applied as droplets from a pipette (or a micro-pipette) or a disposable tip.

The basis for the test is the following:

- A droplet of a liquid of lower surface tension will *spread* upon a surface whose energy level is higher. This lowers the energy of the combined system. The observer using the Nordtest method will see the droplet as spontaneously wetting this surface i.e., a thin film or sheet is formed.

- Another liquid droplet, whose surface tension is higher than that of the surface, will *not spread*. The observer using the Nordtest method will see the droplet as not spontaneously wetting this surface i.e., a small bead is formed. Here "spontaneously" means that the drop will either spread or not within 2 seconds. The observer's first impressions should be considered final.

A surface with relatively high energy will preferentially be spontaneously spread upon by a liquid with a relatively lower surface tension, thus decreasing the overall energy of the system. In other words, if a liquid has more attraction for a solid than it does for itself, it will spread on the solid. The liquid will not spread on the solid if it has more attraction for itself than for the solid.

Different materials, when clean, have different values of surface energy. Values from reference 23 for different materials are listed in Table 1. Since cleanliness is relative, the values in Table 1 should be taken only as a guide. Baseline values for a clean surface should be determined locally. Here, clean refers to a surface free of molecular contamination such as oil, grease, wax, fingerprints, pyrolized residue, or an aqueous solution. Surface energies of metal surfaces free of atomic contamination such as oxide layers are considerably higher than the values in Table 1.

The test liquids used in the Nordtest Poly 176 are mixtures of ethanol and water, *though other combinations can be used*. Ethanol and water both naturally disappear from the surface under test (evaporate), are inexpensive, and are harmless to people,

Table 1.
Surface energies used in Nordtest Poly 176 for clean surfaces [23]

Material surface	Surface energy mJ/m^2
Hot rolled steel	32 to 34
Cold rolled steel	30 to 32
Stainless steel	26 to 28
Aluminum	24 to 26
Bronze	24 to 26
Hot rolled steel	32 to 34
Cold rolled steel	30 to 32
Stainless steel	26 to 28
Aluminum	24 to 26
Bronze	24 to 26
Copper	22 to 24
Polyethylene	22 to 24

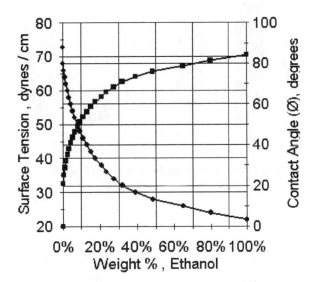

Figure 3. Calibration of test liquids for the Nordtest Poly 176 test method (contact angles are on steel).

the environment, and most substrates. Other binary mixtures could be chosen. A calibration curve is plotted as Fig. 3 [24].

No equipment is involved other than the pipette, which does not have to apply a controlled volume. The test result is the conformation of that volume when applied to a surface.

4.2.1. Four case histories with use of the Nordtest Poly 176 method
Reference 23 contains limited application data, but eight case histories are reviewed.
Four are summarized below. Measurements of surface energy were used for the
following:

● To troubleshoot an operation where threads on a bolt and a mating threaded gland
  were bonded together. An acid etching process increased the metal surface energy
  from 30 to 50 mJ/m^2 which was just above that of the adhesive at 46 mJ/m^2. The
  surface energy of the adhesive was higher than that of the un-etched surface and
  would not wet it. See Fig. 4.

● To identify why there was periodic loss of adhesion between a topcoat and a
  primer. The primer had a value of about 23 mJ/m^2 which nearly always exceeded
  that of the topcoat. For good wetting, the surface energy of the topcoat needed to
  be less than that of the primer. See Fig. 5.

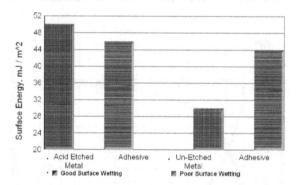

Figure 4. Surface energy measurements using Nordtest Poly 176 of metal with and without acid
treatment.

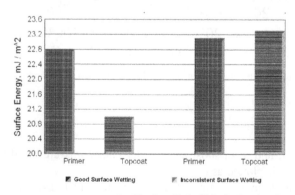

Figure 5. Surface energy measurements using Nordtest Poly 176 of primed and topcoated metal.

- To show zinc chromate treatment of a metal surface was not applied correctly. A treatment step with phosphoric acid had been omitted. That step was necessary to elevate the metal surface energy to at least 30 to 34 mJ/m^2 so that an adhesive with a lower surface energy could consistently bond to it. See Fig. 6.

- To identify retained solvent on a metal surface ($\sim$24 mJ/m^2) as an apparently unavoidable contaminant in the application of a high solids coating. The coating was reformulated with a solvent having lower surface tension (n-butanol at 28 mJ/m^2) vs the previous one (xylene at 24 mJ/m^2). Adhesion improved because the surface tension of the retained coating solvent was slightly less than the surface energy of the metal surface. See Fig. 7.

In summary, the Nordtest Poly 176 allowed solution of surface cleaning problems without major investment for facilities or training.

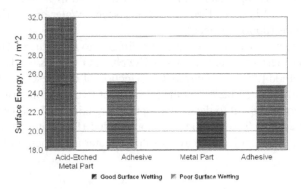

Figure 6. Surface energy measurements using Nordtest Poly 176 of zinc before and after acid treatment.

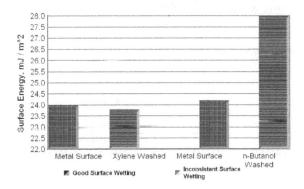

Figure 7. Surface energy measurements using Nordtest Poly 176 of washed and unwashed metal.

5. THE WATER-BREAK TEST

The basis for the water-break test, referred to earlier by Devaux [7] and Nelson [9], is simple:

- If water "beads," the surface is considered to be contaminated with a hydrophobic substance (oil/grease).

- If the water "breaks" or "sheets" the surface is considered clean.

In Fig. 8 the contact angle is very high which is a characteristic of a poorly cleaned surface. In Fig. 9 the contact angle is very low which is a characteristic of a well cleaned surface.

An actual demonstration of an outcome of the "water-break" test is shown in Fig. 10. The image is of Rhodium sheet that has been well cleaned (left) and poorly (right) cleaned by a proprietary solution of sulfuric acid in water. Water on the left piece (well cleaned) is in the form of a film or sheet. Water on the right piece is in the form of beads (poorly cleaned).

The "water-break" test is subjective and difficult to reproduce. The outcome of a "water-break" test is usually binary – GO/NO GO (sheets/beads). There is an ASTM test method [25] which covers the detection of the presence of hydrophobic (nonwetting) films on surfaces and the presence of hydrophobic organic materials. The test will not be useful to detect hydrophilic organics such as wetting agents or inorganic impurities. The method calls for purified (DI) water to be used, rather than tap water. This choice is to avoid soluble materials which will affect surface tension.

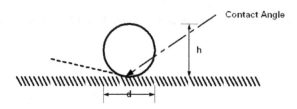

Figure 8. High contact angle of droplet on a surface.

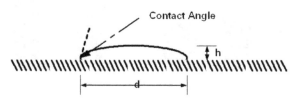

Figure 9. Low contact angle of droplet on a surface.

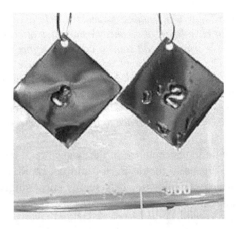

Figure 10. Extremes of results from water-break test – good wetting (left), poor wetting (right).

5.1. Opinions about the water-break test from experienced users

This test still maintains credibility in the electronics, metal bonding, and metal manufacturing areas. Some quoted published opinions about it should suffice to describe the value it generally brings in cleanliness evaluations.

This opinion is from a posting on TECHNET by Brian Ellis to Ranier Blomberg, March 1, 2000:

> *I'm sorry, but the so-called water-break test is as useless as the name is long. It is one of the biggest myths and fallacies that occur in our industry and should be drowned in the water required for it and then laid to permanent rest. ...Can we please lay this old red herring to its well deserved and everlasting rest?*

This opinion is also by the author noted above [26]:

> *Technical managers of fabricators using the test may wish to re-examine whether it is valid under their particular circumstances. If it were replaced by a more scientific test covering their particular needs, it is possible that the reliability of subsequent operations may be improved, increasing production yield.*

This opinion is quoted from an FAA report about assembly of aircraft structures [27]:

> *The most common methods of monitoring the surface preparation were visual checks, water-break tests, witness panels and surface chemistry tests.*

This opinion is from an industrial manual of instructions for metal plating [28]:

The simple water-break test remains the most widely used method, in shop practice, to test for cleanliness. A cleaned metal panel, after being submerged in cold, clean, clear water, would show a continuous film of water over the entire surface. The presence of any residues will cause the water film to break, leaving those areas not wetted. Sometimes small areas will fail to show up immediately with this test, so sufficient time should be allowed to ensure that the film is continuous over the entire surface. A Surface Quality Meter (contact angle measurement) is needed to replace the water-break test.

In summary, the continued acceptance of the water-break test (with the acknowledged flaws it brings) attests to the desire for simple, rapid, and low-cost systems for evaluation of surface cleanliness. This seems surprisingly true in high-value applications such as airframe construction.

6. FOUNDATION OF WETTABILITY MEASUREMENTS

It was Thomas Young, M.D. who, in 1805, quantified the observation about a liquid drop. Equation (1), named for him, relates the shape of a liquid drop resting on a surface to three forces (tensions) which act on it.

$$\gamma_{SV} - \gamma_{SL} = \gamma_{LV} \cos \theta \qquad (1)$$

The parameter γ is a surface energy. Surface energy describes reactivity of solid surfaces. The units of surface energy are free energy per unit area (mJ/m^2). The units of surface tension are force per unit length (dyne/cm or mN/m). A numerical value of mJ/m^2 can be converted via the definitions of these units to the same numerical value of dyne/cm.

But surface energy is not the same as surface tension. Surface tension results from intermolecular forces between molecules. At a surface, molecules of a liquid undergo a net force.

In Equation (1), θ is the contact angle, measured in degrees (or radians). It is the independently measured parameter. The subscripts represent the three interfaces and are: SV = solid–vapor, SL = solid–liquid, and LV = liquid–vapor. In general: γ_{LV} is known as the surface tension between liquid and vapor.

6.1. Limitations of Young's law

Equation (1) applies only to one-dimensional spreading in the horizontal direction. Young's Equation is basically a force balance [29, 30]. The force of buoyancy, gravity, is neglected in this force balance because the direction of summation of forces is horizontal. Equation (1) can also be derived through minimization of the total surface free energy.

A second, and unstated, assumption in Young's equation is metastable equilibrium. This assumption is often sound as the rates of drop movement are low but certain. The sum of all the tensions acting at the edge of the drop must balance to zero for equilibrium to hold.

Two extremes are shown in Fig. 8 (a drop of liquid which is repelled by the solid surface) and Fig. 9 (a drop of liquid which wets the solid surface).

For the same liquid volume, the conformation of the liquid drop is very different depending on whether the liquid is compatible with the surface, or not.

- In Fig. 8, the liquid is repelled by the surface. The volume of the liquid is shaped so that surface energy is minimized and the height of the liquid volume is a maximum (large). The contact angle is very high. In the extreme of total repulsion of the liquid by the surface, the contact angle would be 180 degrees.

- In Fig. 9, the liquid wets the surface. The volume of liquid is spread out, and the height of the liquid volume is low (small). The contact angle is very low. It would be zero if the surface and the liquid were perfectly compatible.

Young's equation applies to ideal sections of surfaces that are perfectly smooth and chemically homogeneous (not necessarily clean, just uniformly soiled within the chosen section).

In only a very few cases is the information about the cleanliness of the surface obtained directly from Young's Equation. Usually some modified form is solved as additional equations and constraints are added to make solution feasible. Some of these additional equations and constraints are described in the next section, and in a separate reference [31].

6.2. Derivations based on Young's equation

There are four terms in Young's equation. Only the two on the right (θ and γ_{LV}), the contact angle and the liquid–vapor surface tension respectively, are measurable. The latter is conventionally reported in various handbooks as "liquid surface tension" (γ or σ). The other two terms, γ_{SV} and γ_{SL}, are not measurable because a solid has an internal stiffness (which allows it to support a shear stress) so it does not deform as does a liquid.

Their difference ($\gamma_{SV} - \gamma_{SL}$) can be calculated by Young's equation (because θ and γ_{LV} are known), but individually they cannot be evaluated solely with that expression. In other words, there are two unknowns (γ_{SV} and γ_{SL}) and one equation (Young's).

This dilemma is conventionally unraveled by assumption of some relationship between the two unknown tensions, or adoption of another assumption. In this way they can be combined into a single one, or independently evaluated.

Table 2 contains descriptions of a few of these assumptions about γ_{SV} and γ_{SL} and their results. This is by no means a complete listing [32], but those noted here have proven of value in surface cleaning work. For example, the Owens–Wendt Equation is not included because its use requires measurement of contact angles with two liquids, one of which is seldom found in shops doing cleaning work.

Table 2.
Further derivations from Young's equation useful in evaluation of surface cleanliness

Developer(s)	Assumption	Equation	Comment
–	$(\gamma_{SV} - \gamma_{SL})$ is relatively constant for a surface	$\gamma_{SV} - \gamma_{SL} = \gamma_{LV}\cos\theta$ (2)	–
Zisman [33]	Each surface has a single wetting tension (surface energy). $\cos\theta$ for many liquids *on the same surface* are plotted vs γ_{LV}. γ_C is extrapolated from that plot to where $\cos\theta = 1$ [$\theta = 0°$ or perfect wetting]	$\gamma_C = \gamma_{LV}\cos\theta$ (3)	The assumed linear relationship between $\cos\theta$ and γ_{LV} is mostly valid where surface wetting is poor. γ_C is the critical surface tension of wetting
Young–Dupre	The work of adhesion (W_A) between a liquid and a solid can be measured	$W_A = \gamma_{LV}(1 + \cos\theta)$ (4)	More of a starting point for subsequent calculations than being useful for evaluation of a surface condition
Girifalco–Good Fowkes–Young (GGFY) [34]	Work of adhesion can be calculated as the geometric mean between γ_{SV} and γ_{LV} times a proportionality constant [35] which is a function of the molar volumes of components	$\cos\theta = -1 + \frac{2\sqrt{\gamma_{SV}\cdot\gamma_{LV}}}{\gamma_{LV}}$ (5)	Most useful rule for manufacturing work and is used in this paper. γ_{SV} is a characteristic of the solid, which may be taken as proxy for cleanliness

These derivations allow measurements of contact angle to be converted into estimates of surface character (energy). As stated earlier, these estimates of surface character are taken as a proxy for cleanliness.

7. BIKERMAN'S EQUATION

J. J. Bikerman found, more than 60 years ago [36], that for water droplets below a certain size, the contact angle could be found simply by measuring the droplet diameter, if the drop volume was known [37]. One uses a micro-pipette or micro-syringe to dispense a selected volume from 1 to 20 microliters [38]. This can be the same equipment which is used to inject liquid samples in gas chromatography (GC). See Fig. 11.

This technique involves estimation of contact angle of a tiny and known volume liquid (usually water) via measurement of its diameter. Equation (6) [39] relating

Figure 11. Syringe for application of liquid droplets.

Figure 12. Water droplet on grated surface.

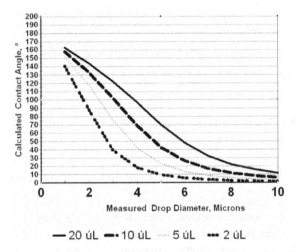

Figure 13. Solution of Bikerman equation (6) for larger contact angles (less clean surfaces).

controlled drop volume, measured diameter, and calculated contact angle (θ) is:

$$\frac{[Drop\ Diameter]^3}{Drop\ Volume} = \frac{24 \times \{\sin\theta\}^2}{\pi \times (2 - \{3 \times \cos\theta\} + \{\cos\theta\}^3)} \quad (6)$$

The best way to measure drop size is with magnification, *viewed from the top*. See Fig. 12. A digital camera can record the image within one or two seconds after the drop is deposited on the surface by the syringe. Then the digital image can be examined by whatever software is available, and the diameter estimated.

Facilities needed here (a pipette or syringe, a magnifier, and a vernier scale) should be called negligible in the budget of any organization which values cleanliness.

Equation (6) is not solved by direct algebra. Iterative solutions are shown in Fig. 13 and Fig. 14 for drop sizes of 20, 10, 5, and 2 microliters. At the 20 microliter size, gravitational forces may distort the one-dimensional description of Equation

J. B. Durkee and A. Kuhn

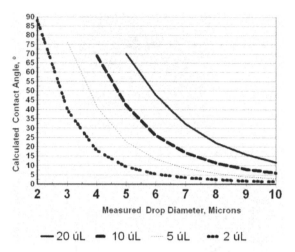

Figure 14. Solution of Bikerman equation (6) for smaller contact angles (more clean surfaces).

(1). Figure 13 is for larger contact angles (less clean surfaces). Figure 14 is for smaller contact angles (more clean surfaces).

7.1. Methodology for use of the Bikerman equation to manage cleanliness

Bikerman's ideas seem to have lain fallow until the 1970s, when Miller [40], at Lockheed, revived them and demonstrated their use in a major aircraft paint shop where fuselages and other components were cleaned and painted. Reference 40 contains a conspicuous and diverse collection of data about evaluations of the quality of cleanliness of various sections of the Lockheed C-130 aircraft assemblies.

Miller used a low-power optical microscope with calibrated reticle to measure drop diameter and then determined the contact angle by reading the diameter from a graph similar to Fig. 13 or 14, or a series of nomograms. Replication data are included in the reference, and appear satisfactory. Miller also used Zisman's approach, Equation (3) from Table 2, to estimate surface energy from the measurement of contact angles. Instead of using multiple data points of Cos θ vs γ_{LV}, a linear relationship was assumed by Miller based on only two measurements.

We tested the validity of Miller's 2-point calibration procedure for the liquid soils and metals of significance to Lockheed, and converted Miller's published raw data to surface energy using the GGFY Equation (5) from Table 2.

$$\cos \theta = -1 + \frac{2\sqrt{\gamma_{SV} \cdot \gamma_{LV}}}{\gamma_{LV}} \tag{5}$$

The comparison of calculations is shown in Fig. 15. The excellent agreement validates Miller's procedure.

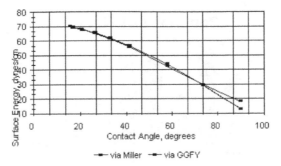

Figure 15. Comparison of experimental measurements of surface energy and contact angle with values predicted from GGFY equation (5).

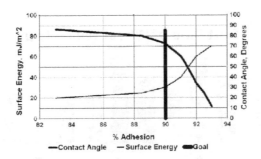

Figure 16. Calibration of contact angle and surface energy measurements against desired adhesion.

7.2. Case histories of use of the Bikerman equation to manage cleanliness

Miller began by developing a screening procedure to identify the minimum level of surface energy necessary to meet the specification for 90% paint adhesion.

Chromate-clad aluminum panels were cleaned and then deliberately soiled using various quantities of stearic acid, to create a set of samples with surface energies ranging from 10 to 70 mJ/m^2 (equivalent to dyne/cm in Fig. 15). These panels were then coated with the US Navy epoxy-polyamide paint system, and after curing, the adhesion of the paint was tested. Results are shown in Fig. 16. The minimum acceptable surface energy was found to be about 30 mJ/m^2 (dyne/cm) to achieve 90% adhesion. The measured contact angle was 72.8 degrees.

Three of the many examples of the evaluations conducted by Miller are shown below in Figs. 17 to 19. Various portions of an airframe were cleaned with various methods in multiple applications. Then various airframe locations within each portion were tested for cleanliness before paint was applied.

- In Fig. 17, it is clear that for all three test locations on the port side tailplane area, a second cleaning step was unnecessary and represented needless effort and cost.

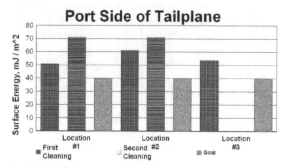

Figure 17. Surface energy values calculated from local measurements of droplet diameter and calculation of contact angle through the Bikerman equation (6) and the GGFY equation (5).

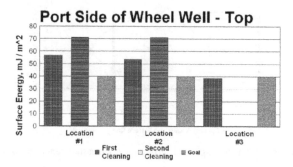

Figure 18. Surface energy values calculated from local measurements of droplet diameter and calculation of contact angle through the Bikerman equation (6) and the GGFY equation (5).

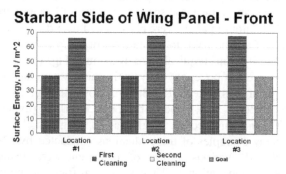

Figure 19. Surface energy values calculated from local measurements of droplet diameter and calculation of contact angle through the Bikerman equation (6) and the GGFY equation (5).

- In Fig. 18, it is clear that for all three test locations on the port side wheel well-top area, the first cleaning step was not adequate and a second step of cleaning treatment was necessary to meet paint adhesion goals.
- In Fig. 19, the same point is made as in Fig. 18, and reproducibility demonstrated at all three locations of the front part of the starboard side wing panel.

Four aspects of Miller's work are unique:

- The parts being cleaned were large (airframe components), as opposed to small components such as connectors or lenses, which are the type of parts upon which wettability measurements are normally made.
- The part surface being tested was too large to fit within a goniometer. Since the surface being evaluated was large, workers evaluated various portions of it in place by measuring the diameter of applied drops.
- There is a steep dependence, as shown in Fig. 16, between surface energy or contact angle and percent adhesion. The goal value of 90% adhesion to aluminum is achieved at a calculated surface energy of 30 mJ/m^2. But the calculated surface energy must increase to about 70 mJ/m^2 for the percent adhesion to increase to 93%. The calculation of surface energy uses Equation (5) and Miller's calculated values of contact angle based on measurements of droplet diameter and the Bikerman Equation (6).
- The liability for poor cleaning was high because the assembled airframes engendered a high level of scrutiny by government regulators.

Miller plainly showed that in high-liability situations, legitimate and valuable cleaning work can be tested without complex and expensive facilities.

Lockheed management were so impressed with the utility, simplicity, and cost of this method of cleanliness management, they believed other firms would be attracted by those values. In the 1970s, Lockheed sold a kit of a microscope and the associated calibration graphs under the trade name "Surf-Scope." The product is no longer commercially offered.

8. LOW-COST EQUIPMENT FOR WETTABILITY MEASUREMENTS

Accurate measurement of the contact angle is not an easy task – with any equipment. Reproducibility usually requires scrupulously clean experimental conditions and meticulous surface preparation.

Simple equipment that has produced useful results (though not associated with surface cleaning) is the reflecting goniometer. "With some training, the direct measurement of sessile-drop contact angles with a telescope and a goniometer eyepiece can rapidly and inexpensively yield results to within approximately 2 degrees" [41]. Today, the US and global market for goniometers is served with units whose purchase price is closer to US $10,000 than $100. Costs of this magnitude are hard to justify for management of common cleaning tasks. These authors believe

Figure 20. Low cost goniometers for cleanliness evaluation of pieces with a small surface area.

this is a major reason the unreliable water-break test has so much currency of use today.

Two examples of commercial goniometers which can be justified for most surface cleaning operations are shown in Fig. 20; each costs less than $500. A user rotates a focused beam of light about a liquid drop and carefully notes the angle at which reflected light from the drop appears or disappears. The model in the foreground has a view screen for examining the conformation of the droplet with a vernier. The model in the background is intended to be coupled to a digital camera. Both are portable and powered by common batteries. Claimed accuracy of the former model is ±2 degrees. For the latter, the claim is ±5 degrees, which may disqualify it for managing surface cleanliness. Obviously, neither model can be used with large samples, in which the Bikerman Equation (6) can be used.

8.1. Negative case history of use of low-cost equipment to manage cleanliness

Uhlman [42] reports about the wettability of metal surfaces after blasting with dry ice crystals. In the work, a proprietary protractor type goniometer was used to measure contact angles.

The results in Fig. 21 are interesting because they confirm how application of blast media has different consequences than does application of other cleaning products. Application of blast media blasting can produce clean surfaces. But rough surfaces are usually also produced if too much blast medium, or the wrong type, is applied for too long a time. In Fig. 21:

- Blasting at the highest rate (100 kg/hr) first cleans the soil from the surface, then roughens the surface as shown by the initial decline in contact angle and an increase of contact angle with further blasting exposure.
- Blasting at the lowest rate (50 kg/hr) never cleans the soil from the surface. This is shown by the contact angle never declining below the initial value.

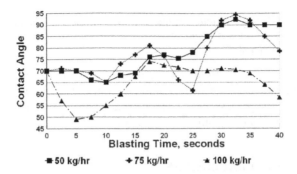

Figure 21. Contact angle measurements on steel surfaces cleaned with blast material.

• Additional blast treatment time can produce a less rough surface via degrading peaks above the mean surface configuration. This produces a reduced measurement of contact angle.

Wettability measurements, unlike most cleanliness tests, do not differentiate between changes in surface cleanliness and roughness. That is why blast cleaning is a method of cleaning where low cost (or high cost) wettability measurements should not be used to manage surface cleanliness.

9. CHOICE OF WETTING LIQUID

Which liquid should be used for testing to establish that surface contamination exists and to estimate its level? The de facto choice is water. Should water always be used?

The choice of liquid does matter: "contact angle has no meaning without specifying the test liquid. ... Also, omitting [specification of] the test liquid would be like taking the labels off the dyne solution pens. You really would not know much" [43].

The test liquid for *all* cleanliness testing (the "water-break" test, dyne liquid tests, Nordtest Poly 176, etc.) should be chosen according to the nature of the soil which is contaminating the surface.

The choice should *always favor a liquid which is repelled by the soil.*

By this choice, when the surface is completely free of contaminant which repels the test liquid, the test liquid will fully wet the cleaned surface (to the extent the test liquid is compatible with that surface). If the soil and the test liquid were compatible (miscible), the test liquid would wet the soiled surface, leading an observer to believe the surface was clean when it was not so.

Here are some suggestions.

• For an aqueous soil such as a water-based cutting fluid, choose a hydrocarbon test liquid such as any paraffinic solvent.

• For an oil or oil-like soil such as a lubricating liquid, choose water.

If one would choose a cleaning agent to be most compatible with the soil being cleaned, the wetting test liquid should be chosen to be least compatible with the soil being cleaned.

This choice is made because of the focus of this paper: surface cleanliness, not surface characteristic. An excellent example is testing for completion of rinsing in an aqueous cleaning process. Application of a drop of water to a surface wetted with dirty water will produce no new information.

10. SUMMARY AND CONCLUSIONS

It was discovered that there is significant demand by those managing cleaning operations for a simple, effective, and low-cost method for measurement of surface cleanliness. There is significant dissatisfaction with the water-break test.

Low-cost approaches have been proven to enable cleanliness management through the measurement of surface energy or contact angle. Contact angle can be converted to equivalent surface energy through various theoretical analyses of surface forces.

Data provided in this paper about use of such approaches in cleaning management [44] include:

- Use of surface tension test liquids (dyne liquids). Both commercial ones (Fig. 1) and user-prepared ones (Fig. 3) can be employed, though with different procedures. The outcome in either case is an estimate of the surface energy of the cleaned or uncleaned surface. Performance data in cleaning management are shown in Fig. 2, and Figs. 4 through 7.
- Direct measurement of the contact angle between a drop of chosen liquid and the surface being evaluated. One can do that with reflecting-light goniometer such as that in Fig. 20. Performance data in cleaning management are shown in Fig. 2.
- Indirect calculation of the contact angle by carefully measuring the normally projected diameter of the drop and calculate the contact angle through principles of solid geometry, as in the Bikerman Equation (6). Performance data in a critical area of cleaning management are shown in Figs. 17 through 19.
- Identification that wettability measurements are not a good choice for management of blast cleaning. Here the surface may be both cleaned and roughened. Measurements of contact angle cannot discriminate between those two outcomes.

There is a proven compromise in managing surface cleanliness through measurement of surface wettability. One does not have to accept the imprecision and irreproducibility of the water-break test or the expenditure of funds for sophisticated facilities.

REFERENCES

1. J. B. Durkee, *Management of Industrial Cleaning Technology and Processes,* Chapter 5, pp 257-294, Elsevier, Amsterdam (2006).

2. K. L. Mittal (Ed.), *Contact Angle, Wettability and Adhesion,* Vol. 3, VSP/Brill, Leiden (2003).
3. J. C. Harris, *ASTM Special Technical Publication,* Vol. 90 (1953).
4. L. Moser, *Annalen Physik Chemie,* **134,** 105-111 (1843).
5. L. Moser, *Annalen Physik Chemie,* **136,** 40-48 (1843).
6. A. Pockels, *Annalen Physik Chemie,* **313,** 854-871 (1902).
7. M. H. Devaux, *Physique Radium,* **4,** 293-309 (1923).
8. E. M. Baker and R. Schneidewind, *Trans. Electrochemical Soc.,* **45,** 327-352 (1924).
9. C. Nelson, *Steel,* **113,** 106-108, 132-136 (1943).
10. K. L. Mittal (Ed.), *Contact Angle, Wettability and Adhesion,* Vol. 4, VSP/Brill, Leiden (2006).
11. L. Zakarias, *Chime Industrie,* **36,** 1095-1100 (1936).
12. K. L. Mittal (Ed.), *Surface Contamination – Genesis, Detection, and Control,* Vols. 1 and 2, Plenum, New York (1979).
13. See http://www.pillartech.com/corona_tech3a.html for additional information.
14. E. Boyle, *Paper, Film, & Foil Converters,* (September 1, 1996). Article available at http://pffc-online.com/mag/paper_taking_measure_surface/
15. A. T. Kuhn, *Metal Finishing,* **103,** 72-79 (May 2005).
16. Three suppliers of these products can be found at http://www.shermantreaters.co.uk, http://www.tigres.de/, and http://www.softal3dt.com
    As a product, they are produced in 1 or 2 mN /m (dyne/cm) increments. A container of them resembles a box of crayons.
17. C. M. Hansen, *J. Paint Technology,* **42,** 550, 660-664 (1970).
18. ASTM D2578-04a, "Standard Test Method for Wetting Tension of Polyethylene and Polypropylene Films, 2004." This method uses mixtures of ethylene glycol monoethyl ether and formamide which are not readily volatile and may be good solvents for some soils. The latter is decidedly not wanted in a cleanliness test liquid.
19. ISO 8296, "Plastics – Film and Sheeting – Determination of Wetting Tension," (2003).
20. JIS K6768, "Polyethylene and Polypropylene Films," (1971). JIS K6768 was recently revised and is now substantially equivalent to the ISO and ASTM Standards.
21. DIN 53 364, "Plastics - Film and Sheeting - Determination of Wetting Tension," (2003).
22. M. Stiles, T. Hausner and B. Haase, *J. Oberflaechentechnik,* 66-71 (June 1998).
23. C. M. Hansen, *Pigment Resin Technol,* **27,** 304-307 (1998). Nordtest Poly 176 can be obtained at http://www.nordicinnovation.net/nordtestfiler/poly176.pdf
24. The unit for surface tension in Figure 3 is dyne/cm, which is equivalent to milli-Newtons/meter (mN/ m).
25. ASTM F22-02, "Standard Test Method for Hydrophobic Surface Films by the Water-Break Test."
26. B. Ellis, *Circuit World,* **31**(4), 47-50 (2005).
27. "Assessment of Industry Practices for Aircraft-bonded Supports and Structures," *FAA Report on Bonded Structures,* presented at a workshop at Wichita State University (2004).
28. PCS Industrial Sales, "Bright Dipping Aluminum – Industrial Manual" (2005).
29. J. P. Garandet, B. Drevet and N. Eustathopoulos, *Scripta Mater.,* **38,** 1391 (1998).
30. S. W. Ip and J. M. Toguri, *J. Mater. Sci.,* **29,** 688-692 (1994).
31. A complete summary with examples can be found at "Surface Energy Measurements & Cleanliness Testing by Wetting Methods" (www.finishingpublications.com).
32. Two excellent additional summaries are available on the Internet. Authors are R. Woodward (http://www.firsttenangstroms.com/pdfdocs/SurfaceEnergyMethods.pdf) and F. K. Hansen (http://folk.uio.no/fhansen/surface_energy.pdf).
33. W. A. Zisman, in *Contact Angle, Wettability, and Adhesion, Adv. Chem. Ser.,* No. 43, p. 1, American Chemical Society, Washington, D.C. (1964).
34. L. A. Girifalco and R. J. Good, *J. Phys. Chem.,* **61,** 904-909 (1957).
35. The proportionality constant is taken to be 1 for this equation, but has been estimated empirically in the reference: D. Y. Kwok and A. W. Neumann, *Adv. Colloid Interface Sci.,* **81,** 167-249 (1999).

36. J. J. Bikerman, *Trans. Faraday Soc.*, **36**, 412 (1940).
37. J. J. Bikerman, *Ind. Eng. Chem.*, **13**, 443-444 (1941).
38. W. J. Herzberg and J. E. Marian, *J. Colloid Interface Sci.*, **33**, 161-163 (1970).
39. G. L. Mack, *J. Phys. Chem.*, **40**, 159-167 (1936). The author reports his observation that the formula is applicable only for small hemispherical drops (less than 10 μm in diameter) which are not distorted by gravitational effects.
40. R. N. Miller, *Material Protection Performance*, **12**(5), 31-36 (1973).
41. A. W. Neumann and J. K. Spelt (Eds.), *Applied Surface Thermodynamics, Surfactant Science Series Vol. 63*, Marcel Dekker, New York (1996).
42. The reference is located at www.surfacequery.com
    The metal treated was an aluminum-magnesium alloy. Data in Figure 21 are smoothed.
43. Personal communication with Roger Woodward, FirstTenAngstroms, Inc. (April 2006).
44. A CD-ROM e-Book and Toolkit, "Surface Energy Measurements and Cleanliness Testing by Wetting Methods," is detailed at www.finishingpublications.com

Contact Angle, Wettability and Adhesion, Vol. 5, pp. 139–151
Ed. K.L. Mittal
© VSP 2008

Wettability parameters controlling the surface cleanability of stainless steel

LAURENCE BOULANGÉ-PETERMANN,[1,*] JEAN-CHARLES JOUD[2]
and BERNARD BAROUX[2]

[1]*Formerly with Ugine & ALZ, Research Center, BP 15, 62330 Isbergues, France*
[2]*Institut National Polytechnique de Grenoble, Laboratoire de Thermodynamique et Physico-Chimie Métallurgique, 38402 Saint Martin d'Hères Cedex, France*

Abstract—Cleaning of solid surfaces is an important issue in food or catering industry and medical applications. Attention is paid in this paper to the physicochemical parameters controlling cleanability of industrial metal surfaces, taking into consideration the main mechanisms involved in current cleaning processes. Examples are given for some bare and coated stainless steel surfaces.

Oil removal by an aqueous solution containing surfactant is shown to be directly related to the polar component of the metal surface energy: the higher the polar component is, the higher the oil removal. However, instances are given of soiling oil penetrating the surface micro-geometries of metals, which decreases their cleanability irrespective of their polarity. The water contact angle hysteresis is proposed as an overall criterion involving the different aspects of cleanability, at least when surfaces are mechanically cleaned using a liquid flow (e.g. water). Last, a new experimental technique is proposed, aiming to understand better the hydrodynamics of an oil droplet removal (sliding or lifting), opening the route to further modeling of cleaning mechanisms.

Keywords: Stainless steel; cleaning; oil removal; polysiloxane.

1. INTRODUCTION

Stainless steel is a material widely used in the food or catering industry. When the surface of stainless steel comes into contact with nutrients, surface soiling or fouling can occur. It is then necessary to clean it in order to ensure first the surface hygiene as well as to avoid further microbial development and secondly to ensure the equipment performance [1–2] (for instance in heat plate exchangers used in the dairy industry [3]).

To assess the cleanability of a material, different empirical tests may be used, involving natural exposure for an extended time [4], apolar black soiling for

---

*To whom correspondence should be addressed. Tel.: 33 (0) 476 689 421; Fax: 33 (0) 476 683 773; e-mail: laurence_boulange@europe.bd.com

Current address: BDMedical Pharmaceutical Systems, 11 rue Aristide Bergès, 38800 Le Pont de Claix, France.

Table 1.
Surface free energy components (mJ/m^2) derived from water (θ_W), formamide (θ_F) and di-iodomethane (θ_D) contact angles (degrees) [12]. The standard deviation in contact angles is given in parentheses

Samples	θ_W	θ_F	θ_D	γ_S^d	γ_S^p
Bare surfaces					
Ss$_1$ (bright anneaedl)	70 (5)	57 (3)	60 (3)	27	11
Ss$_2$ (pickled)	43 (3)	42 (2)	49 (3)	30	25
Ss$_3$ (textured)	49 (3)	37 (3)	44 (3)	34	20
Ss$_4$ (textured)	44 (2)	41 (2)	57 (3)	27	27
Ss$_5$ (chemically attacked)	37 (2)	30 (2)	64 (2)	24	34
Ss$_6$ (mechanically ground)	64 (4)	64 (3)	55 (3)	26	14
Coated surfaces					
Silicon oxide coated Ss$_1$	11 (1)	6 (1)	37 (1)	35	36
Silicon oxide coated Ss$_2$	14 (1)	8 (1)	36 (2)	36	35
Polysiloxane coated Ss$_1$	103 (1)	90 (1)	70 (1)	21	1
Polysiloxane coated Ss$_2$	98 (1)	87 (1)	69 (1)	21	2

simulating soiling of buildings in urban environments [5, 6], soiling by oils [7, 8], and removal of pathogenic microorganisms in food industry [3]. On flat steels, the case of cleaning is generally discussed in terms of surface composition [9] topography or polarity [10]. In this work, we investigate the cleanability of bare and coated materials under controlled flow geometry.

The aim of this study was to define the main surface parameters (e.g. topography, surface energy) playing a determining role in the cleaning mechanisms of soiled industrial surfaces. For this purpose, six different bare surfaces obtained by varying the final stage of the steelmaking process on the same metal substrate (SS30400 Unified Number System, 1 mm thick stainless steel sheets) were first considered and referred to as Ss$_1$ to Ss$_6$ (Table 1).

2. MATERIALS AND METHODS

2.1. Selection of materials

These surfaces differ in the the final treatment. Bright annealed condition (Ss$_1$) means that the final annealing of cold rolled sheets is performed in a hydrogen containing atmosphere and does not need any subsequent chemical pickling. However, the water content in the atmosphere is sufficient to form a passive film [11]. On the contrary, when this annealing is performed in an oxidizing atmosphere, a final pickling is carried out in order to remove the oxide scale (Ss_2) formed during annealing. Then, the film formation is completed by rinsing with water and further exposure to an ambient atmosphere (relative humidity of 30%). Textured surfaces (Ss$_3$ and Ss$_4$) were obtained using textured rolls. The chemically etched sample (Ss_5) was produced from an initial Ss_2 sheet by immersing for 10 min in a 60% HNO_3 solution at a current density of 120 mA/cm^2. Last, a mechanically ground sample (Ss_6) was obtained from an initial (Ss_2) sheet processed in laboratory conditions using abrasive strips with different sizes of carbide particles (120 and 320 μm).

Thus, in order to drastically modify the surface hydrophobicity without altering significantly the surface topography, different coatings were deposited: polysiloxane and silicon oxide were deposited by plasma assisted chemical vapor deposition (PACVD) with a mixture of O_2 and hexamethyldisiloxane (HMDSO) in the reactor chamber. The detailed coating conditions were given in a previous paper [12].

2.2. Surface free energy

On both bare and coated surfaces, contact angles (θ) were measured using a series of three pure liquids (water, formamide and diiomethane) with the sessile drop technique. The dispersion (γ^d) and polar (γ^p) components of the solid (S) surface free energy were calculated by combining the Young, and Owens and Wendt's equations:

$$\gamma_{LV} \frac{\cos\theta + 1}{2\sqrt{\gamma_S^d}} = \sqrt{\gamma_S^p}\sqrt{\frac{\gamma_L^p}{\gamma_L^d}} + \sqrt{\gamma_S^d} \tag{1}$$

where γ_S^d and γ_S^p represent, respectively, the dispersion and polar components of the solid surface free energy and γ_L^d and γ_L^p represent, respectively, the dispersion and polar components of the liquid surface tension which are known.

2.3. Surface roughness

The selected parameters were the arithmetic average roughness (R_a) and the maximum peak-to-valley height (R_t) expressed in μm. These parameters were deduced from an optical profilometry (Microsurf 3D, Fogale, Montpellier, France) (scans of area 100×100 μm^2) using the Surfvision software.

2.4. Cleanability test

The cleaning performance was assessed using a laminar flow cell where the sample to be cleaned was placed at the bottom of the cell. Prior to cleaning, the surface was soiled by an oil droplet which was mixed with a dye to discriminate the soil from the solid. Previous papers described in detail the method used to assess the cleaning performance of materials [13, 14].

3. ROLE OF THE SURFACE POLARITY

From a physicochemical point of view, one should keep in mind that bare stainless steel sheets are in fact covered by a nanometric thick passive layer mainly consisting

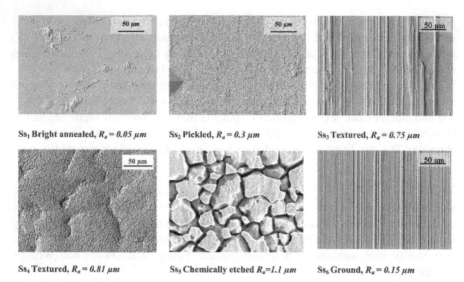

Ss₁ Bright annealed, $R_a = 0.05 \, \mu m$ Ss₂ Pickled, $R_a = 0.3 \, \mu m$ Ss₃ Textured, $R_a = 0.75 \, \mu m$

Ss₄ Textured, $R_a = 0.81 \, \mu m$ Ss₅ Chemically etched $R_a=1.1 \, \mu m$ Ss₆ Ground, $R_a = 0.15 \, \mu m$

Figure 1. Topographic aspects of the bare stainless steel surfaces by scanning electron microscopy [12]. (R_a) is the arithmetic average roughness determined from a 3D profile.

of iron and chromium oxides or hydroxides, that ensures the corrosion protection of the substrate but likely plays an important role in the physicochemical properties of the surface. The 3D views obtained by scanning electron microscopy on these six surfaces are shown in Fig 1. Even on the bright annealed sheet, the surface topography appears complex and some cavities due to this industrial process are seen.

Table 1 shows that water contact angles vary from 37 to 70° on bare stainless steel surfaces. The dispersion component of surface free energy ranges from 24 to 34 mJ/m² but the polar component varies from 11 to 34 mJ/m². After polysiloxane coating, the contact angles are larger and the surface energy is purely dispersive ($\gamma_S \approx \gamma_S^d \approx 20 \, \text{mJ/m}^2$). On the contrary, after silicon oxide coatings, the contact angles are significantly smaller and the polar component of the surface energy is of the same order of magnitude as the dispersion component ($\gamma_S^d \approx \gamma_S^p \approx 35 \, \text{mJ/m}^2$) [13].

On bare as well as coated surfaces (e.g. materials referred to as Ss₁ and Ss₂ and after polysiloxane or silicon oxide coatings), the more polar the surface is, the higher is the oil removal (Fig. 2). This work shows that polar (and hydrophilic) surfaces are favorable for the cleaning after soiling (assuming that the experiment is carried out with similar initial oil soiling) [10, 12] with the creation of solid-cleaning liquid and oil/cleaning liquid interfaces (γ_{SL} and γ_{LO}) and, conversely, the disappearance of solid-oil interface (γ_{SO}):

$$\gamma_{SO} \geqslant \gamma_{OL} + \gamma_{SL} \tag{2}$$

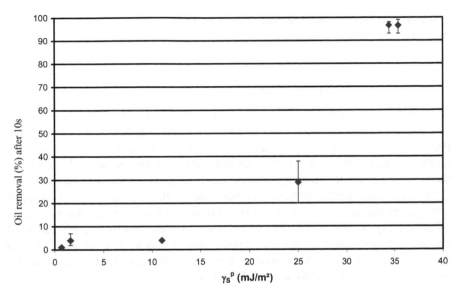

Figure 2. Oil removal after 10 seconds of cleaning in the laminar flow as a function of the polar component of stainless steel [12].

where S, O and L denote, respectively, the solid, oil and cleaning liquid. Combining Eq. (2) with the Owens & Wendt's Eq. (1) results in:

$$2\left(\sqrt{\gamma_L^p \gamma_S^p} + \sqrt{\gamma_S^d}\left(\sqrt{\gamma_L^d} - \sqrt{\gamma_O^d}\right) + \sqrt{\gamma_O^d \gamma_L^d} - \gamma_L\right) \geqslant 0 \qquad (3)$$

Eq. (3) can be split into a positive term (E_r) depending on the polar component of the solid surface energy (γ_S^p) and a negative term (E_a) which is a function of the dispersion component of the solid surface energy (γ_S^d); r and a denote, respectively, the removal or the adhesion of soil on the surface. Finally,

$$E_r + E_a > 0 \qquad (4)$$

where

$$E_r = 2\left(\sqrt{\gamma_L^p}\sqrt{\gamma_S^p} + \sqrt{\gamma_O^d \gamma_L^d} - \gamma_L\right) \qquad (5)$$

and

$$E_a = 2\left(\sqrt{\gamma_S^d}\left(\sqrt{\gamma_L^d} - \sqrt{\gamma_O^d}\right)\right) \qquad (6)$$

To explain the difference between hydrophobic polysiloxane and hydrophilic silicon oxide coatings, we propose that the surfactant molecules orient their polar heads towards the solid surface. This enhances their adsorption onto the polar surfaces. The surfactant molecules can slide under the oil droplets and thereby easily remove them.

4. HIGHLY ROUGHENED SURFACES

In addition to the surface polarity, the surface micro-geometry also plays a major role in cleanability. This effect was assessed by comparing two batches of surfaces ((Ss_1, Ss_3 and Ss_6) with (Ss_2, Ss_4 and Ss_5)). The second batch of samples is much harder to clean than the first [12], even when coated with silicon oxide. Figure 3 show the difference in cleaning performances between Ss_1 and Ss_5 surfaces.

This behaviour is believed to be a consequence of the "impregnation" of a rough substrate by the soiling liquid (oil). Let us consider a rough surface (the roughness factor r being defined as the ratio of the actual surface area to its projected area) consisting of an array of peaks and valleys (respectively in proportions of Φ_s and $1 - \Phi_s$). The condition for the soiling liquid oil to wet not only the peak but also the bottom of the valleys (Wenzel condition) is:

$$\cos\theta_e > \frac{1 - \Phi_s}{r - \Phi_s} \tag{7}$$

with θ_e being the equilibrium contact angle of the liquid on an ideal flat solid surface of the same composition [17]. The measured contact angle θ is given by (this is true

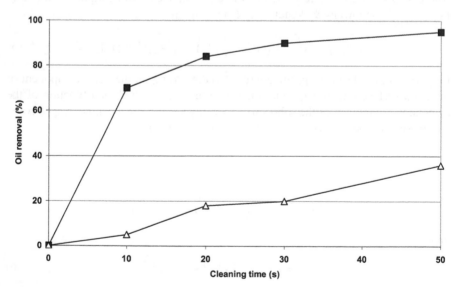

Figure 3. Comparison of the cleaning performances of Ss_1 and Ss_5 surfaces. ■ and △ symbols correspond, respectively, to Ss_1 and Ss_5 materials.

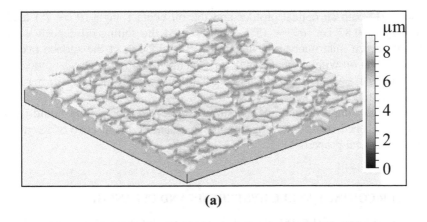

(a)

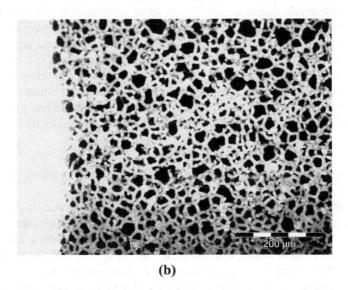

(b)

Figure 4. (a) Surface topography by optical profilometry (scan area 100 μm × 100 μm) with a roughness parameter r of 1.21. (b) Soiling liquid oil impregnation into the solid with a Φ_s value of 0.38. This value was determined after oil drop deposition and staining by image analysis with epifluorescence microscopy.

only when $\theta_e < 90°$):

$$\cos \theta = r \cos \theta_e \qquad (8)$$

showing that a rough surface is more easily wetted than a flat one having the same composition.

This behaviour can be illustrated by a simple experiment (Fig. 4a). We considered a surface (Ss$_5$) displaying large cavities, and first measured its roughness factor

($r = 1.21$) with an optical profiler and the oil contact angle ($\theta = 7°$) leading to $\cos \theta_e = 0.82$ i.e. $\theta_e \sim 35°$. We observed the soiling oil deposit with an epifluorescence microscope and determined the fraction of the surface occupied by the soiling oil with an image analysis system (Fig. 4b), which was found to be 0.38 leading to $\frac{1-\Phi_s}{r-\Phi_s} = 0.75$, which is smaller than $\cos \theta_e$, which is characteristic of Wenzel wetting.

Due to this "impregnation" of the surface valleys by the soiling liquid (oil), a highly roughened surface performs poorly in cleaning, irrespective of its intrinsic physicochemical properties.

5. WATER CONTACT ANGLE HYSTERESIS AND CLEANING

As we noted previously [15], the polar component of the surface energy alone is not a sufficient criterion to qualify the cleaning performance of a metallic surface. As demonstrated by the penetration of a soiling liquid into the recesses of a highly roughened surface, an additional parameter is needed to take into account both the surface physicochemistry and a possible impregnation process [15].

Water contact angle hysteresis was measured on different industrial bare and polysiloxane coated surfaces, using the sessile drop method where the volume of the water droplet was increased or decreased until the contact line moved over the solid surface. Contact angle hysteresis (H) is better expressed by its cosine form:

$$H = \cos \theta_r - \cos \theta_a \tag{9}$$

with θ_r and θ_a denoting, respectively, the receding and advancing water contact angles.

After 40 s in the laminar flow cell, cleaning performance was assessed by measuring the percentage of oil removal (R). A poor performance corresponds to R less than 50%, an intermediate one to R values between 50% and 90% and a good one to R values more than 90%. In this case, it is possible to discriminate coated surfaces with poor performance from bare surfaces with intermediate or high performance (Fig. 5). The most easily cleanable surfaces correspond to the most hydrophilic ones with a low water contact angle hysteresis, whereas the least cleanable surfaces are coated and hydrophobic, irrespective of the water contact angle hysteresis. It should be noted that the intermediate group corresponds to less hydrophobic bare surfaces with a higher water contact angle hysteresis. Last, bare materials with good performance or intermediate performance displaying similar contact angle hysteresis ($H = 0.9$) and advancing contact angle ($\cos \theta_a = 0.1$) have an average roughness less than 0.2 μm which is smaller than that of the intermediate group (average roughness R_a more than 0.3 μm R_a denotes the average roughness determined from a 3D profile).

As a general rule, water contact angle hysteresis is a relevant parameter to assess the cleanability of rough surfaces where the soiling liquid can penetrate into the

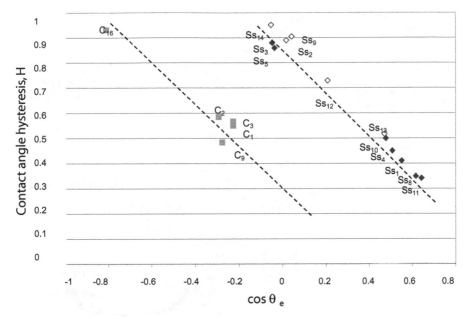

Figure 5. Relation between contact angle hysteresis and $\cos \theta_a$. ♦, ◊ and ■ symbols correspond, respectively, to materials with good cleaning performance where oil removal is more than 90%, intermediate cleaning performance where oil removal is between 50 and 90%, poor cleaning performance where oil removal is less than 50% on bare stainless steels (♦ and ◊) and polysiloxane coatings (■) [15].

solid surface. The most hydrophilic surfaces with the least water contact angle hysteresis are the easiest materials to clean.

6. UNDERSTANDING SOIL (OIL) REMOVAL FROM SOLID SURFACES [16]

For a better understanding of soil removal, we designed a new apparatus combining a laminar flow cell and a goniometer (Fig. 6).

A sliding mode was observed on polysiloxane coatings in the water shear flow with or without the addition of surfactants (Fig. 7a). First, without any shear flow, there is a high affinity between the oil drop and the polysiloxane coating leading to low contact angle. By increasing the shear flow, a critical shear flow is reached, where the hydrodynamic forces overcome the retentive forces leading to oil drop sliding. Finally, no residual oil is left on the surface.

A lift mode was observed on bare stainless steel surface in the water shear flow (Fig. 7b). Without any shear flow, the initial contact angle between the oil drop and the bare surface is high. It should be noted that prior to this experiment, the stainless steel surface was argon plasma treated in order to remove the surface

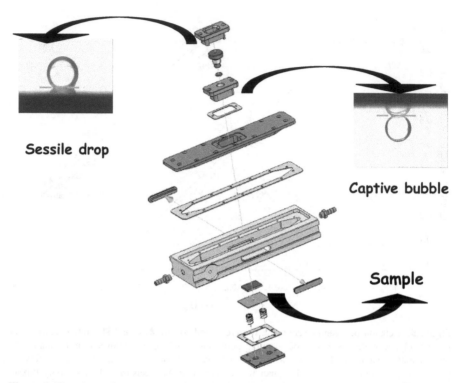

Sessile drop

Captive bubble

Sample

Figure 6. Experimental apparatus coupling a laminar flow cell and a goniometer designed to observe different drop removal modes.

contamination. Then, the shear flow was increased and there is a critical shear flow where the hydrodynamic forces overcome the retentive forces of the oil drop. The drop disintegrates leaving behind a small residue of oil.

A mixed mode was finally observed on the bare surfaces cleaned using the shear flow with surfactants added (Fig. 7c). As previously, the contact angle between the oil drop and the bare surface is high but less than in the previous case, this difference could be explained by the presence of surfactants at the interface. By increasing the shear flow, there is a critical shear flow where the hydrodynamic forces overcome the retentive forces. The oil drop starts to slide and then breaks free from the surface without leaving any residual oil on the solid. A calculation of the critical shear flow (G_c) is possible using a finite elements software (Software package FEMLAB, Comsolab, Stockholm, 2005). There is a decreasing relation between G_c and the volume of the oil drop (Figs. 8a and 8b), irrespective of the cleaning liquid (i.e., water with or without surfactants). In both cases, the G_c is lower on stainless steel surfaces than on polysiloxane coatings. Addition of surfactants in the shear flow

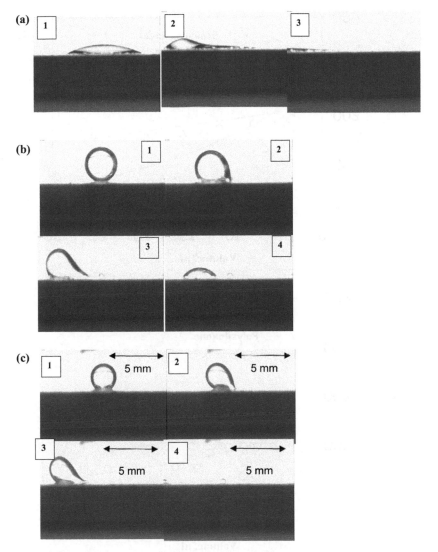

Figure 7. Modes of oil removal. (a) Slide mode observed on polysiloxane coating soiled by oil and cleaned with a water flow with and without surfactants [16]. Picture 1: Soil drop at equilibrium, Picture 2: Drop sliding when a critical shear flow is reached, Picture 3: No residual oil is left on the surface. (b) Lift mode observed on bare surface soiled by oil and cleaned with a water flow [16]. Picture 1. Oil drop at equilibrium, Picture 2: Shear flow is increased and a drop deformation is observed, Picture 3: Hydrodynamic forces overcome the retentive forces, Picture 4: Disintegration of the oil drop and small residual oil on the surface. (c) Mixed mode observed on bare surface soiled by oil and cleaned with a water flow with surfactants [16]. Picture 1: Oil drop at equilibrium, Picture 2: Shear flow is increased and a drop deformation is observed, Picture 3: Drop sliding, Picture 4: no residual oil on the surface.

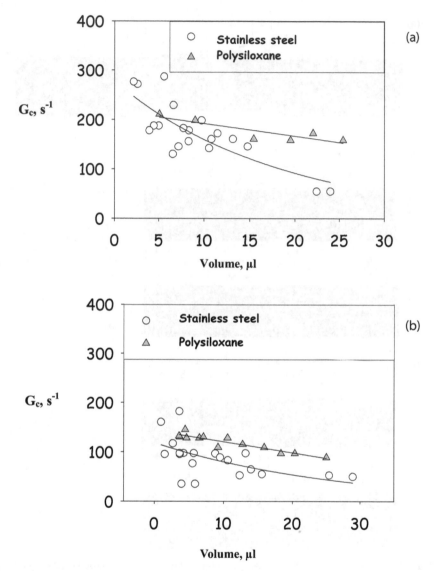

Figure 8. Critical shear flow as a function of the oil drop volume (μL). (a) with a laminar flow of pure water, (b) with a laminar flow of water with surfactants [16].

lowers the G_c values. Last, with a pure water cleaning, it is more difficult to remove very small droplets from stainless steel surfaces.

As a conclusion, there are different modes of oil droplet removal depending on the surface energy of the solid to be cleaned, whether or not surfactants are added, and the hydrodynamic forces induced by the laminar shear flow.

7. CONCLUSIONS

The cleaning performance depends both on the surface energy of the material and its topography. In this respect, a coating can strongly modify the initial properties of bare materials, and also soil affinity and interaction with the cleaning agent. Better understanding of these phenomena will help to develop new materials. From a practical standpoint, there is no universal surface with "good cleaning performance"; each surface has to be adapted to the final application. However, the contact angle hysteresis is a relevant parameter for assessing the general cleanability of a surface.

Acknowledgements

The authors thank Peggy Dusautoir, Jean-Philippe Baron, Philippe Poiret (Arcelor-Ugine-ALZ) for their excellent technical assistance, Audrey Allion and Christelle Gabet (formerly with Arcelor-Ugine-ALZ) and Vanessa Thoreau (INPG-LTPCM) for fruitful discussions, Gregory Berthomé (INPG-LTPCM) for the experimental setup design and Brahim Malki (INPG-LTPCM) for the finite elements calculations.

REFERENCES

1. A. Allion, J. P. Baron and L. Boulangé-Petermann, *Biofouling,* **22**, 269 (2006).
2. M. Demilly, Y. Brechet, F. Bruckert and L. Boulangé, *Colloids Surfaces B,* **51**, 71 (2006).
3. C. Jullien, T. Benezech, V. Lebret and C. Faille, *J. Food Eng.* **56**, 77 (2003).
4. C. M. Rossi, R. M. Esbert, F. Diaz-Pache and F. J. Alonso, *Build. Environ.* **38**, 147 (2003).
5. I. Redsven, R. Kuisma, L. Laitala, E. Pesonen-Leinonen, R. Mahlberg, H. R. Kymäläinen, M. Hautala and A. M. Sjöberg, *Tenside, Surfactants, Detergents* **40**, 346 (2003).
6. H. K. Tikka, M. Suvanto and T. A. Pakkanen, *J. Colloid Interface Sci.* **273**, 388 (2004).
7. R. Foschino, C. Picozzi, E. Giorgi and A. Bontempi, *Annals Microbiol.* **53**, 253 (2003).
8. R. Foschino, C. Picozzi, A. Civardo, M. Bandini and P. Faroldi *J. Food Eng.* **60**, 375 (2003).
9. L. Boulangé-Petermann, C. Jullien, P. E. Dubois, T. Benezech and C. Faille, *Biofouling* **20**, 25 (2004).
10. L. Boulangé-Petermann, C. Debacq, P. Poiret and B. Cromières, in: *Contact Angle, Wettability and Adhesion,* Vol 3, K. L. Mittal (Ed.), pp 501-519. VSP/Brill, Leiden (2003).
11. D. Gorse, J. C. Joud and B. Baroux, *Corrosion Sci.* **33**, 1455 (1992).
12. L. Boulangé-Petermann, C. Gabet and B. Baroux, *Colloids Surfaces A* **272**, 56 (2006).
13. L. Boulangé-Petermann, E. Robine, S. Ritoux, and B. Cromières, *J.Adhesion Sci. Technol.,* **18**, 213-225 (2004).
14. L. Boulangé-Petermann, B. Luquet and M. Valette-Tainturier, in: *Contact Angle, Wettability and Adhesion,* Vol. 4, K. L. Mittal (Ed.) pp. 487-499, VSP/Brill, Leiden (2006).
15. L. Doulange-Petermann, C. Gabet and B. Baroux, *J. Adhesion Sci. Technol.,* **20**, 1463 (2006).
16. V. Thoreau, B. Malki, G. Berthomé, L. Boulangé-Petermann and J. C. Joud, *J. Adhesion Sci. Technol.,* **20**, 1819 (2006).
17. J. Bico, C. Tordeux and D. Quere, *Europhys. Lett.* **55**, 214 (2001).

Contact Angle, Wettability and Adhesion, Vol. 5, pp. 153–171
Ed. K.L. Mittal
© VSP 2008

Functional topcoats on coated textiles for improved or self-attained cleanability

THOMAS BAHNERS,[1,*] LUTZ PRAGER,[2] BÄRBEL MARQUARDT[2] and ECKHARD SCHOLLMEYER[1]

[1]*Deutsches Textilforschungszentrum Nord-West e.V., Adlerstr. 1, 47798 Krefeld, Germany*
[2]*Leibniz-Institut für Oberflächenmodifizierung e.V., Permoserstr. 15, 04318 Leipzig, Germany*

Abstract—The feasibility of UV curable, highly hydrophobic topcoats to improve dirt take-up and cleanability of coated fabrics, as used for the construction of textile roofs, was studied. Besides the chemical composition of the lacquer topcoats, topographic effects were considered, namely the generation of micro-roughness using a novel UV curing process ('micro-folding').

The experimental topcoats had water contact angles on the order of 100 to 115° on smooth surfaces and rather good scratch resistance. In none of the studied systems did the micro-structuring of the surfaces by UV induced micro-folding lead to an increase in hydrophobicity. SEM analyses revealed topographies with peak-to-peak distances S_y of 10 to 20 μm and small aspect ratios, so that it can be concluded that a Cassie–Baxter state of a roughness-dependent contact angle was not achieved in any case.

The experimental topcoats had a reduced take-up of oil- and pigment-containing dirt, as is typical for urban and industrial environments. No correlation was found between the dirt take-up and the water contact angle on the samples, however. It can be assumed from these findings that the mechanisms responsible for the self-cleaning of certain plant and animal surfaces is not necessarily valid for dirt as found in industrial areas with potentially different adhesion to surfaces, take-up by a water droplet, etc. Real-time aging experiments over 7 months were performed and revealed a reduced dirt take-up by all experimental topcoats compared to the conventional system. The dirt take-up was low in the case of the hydrophobic topcoats, but did not decrease further with increasing contact angle. The data did not indicate a positive effect of the micro-rough surface on the extent of dirt take-up.

Keywords: Self-cleaning; coated fabrics; acrylated topcoats; UV curing; micro-folding.

1. INTRODUCTION

Minimized dirt take-up or even self-attained cleanability of technical textiles are important requirements in a number of applications. Scientific literature relates these properties to an extreme hydrophobicity of the surface. Notwithstanding an increasing number of publications over the last few years, the self-cleaning of a variety of animal and plant surfaces, i.e. cleaning by rolling water droplets, has

---

*To whom correspondence should be addressed. Tel.: +49 (0)2151 843-156;
Fax: +49 (0)2151 843-143; e-mail: bahners@dtnw.de

been well-known for many years. Early papers were published in the 1940s, e.g. by Fogg [1] and Cassie and Baxter [2]. It is notable that Cassie and Baxter's work was related to textiles. In later years, this phenomenon had attracted attention e.g. by Abramov [3], Shibuichi *et al.* [4], and Tsujii *et al.* [5] and was publicized as the "Lotus effect" by Barthlott and co-workers (see e.g. [6]).

The wetting behavior of a surface, namely the spreading of a droplet of a given liquid, is basically determined by the relation of the interfacial energies between solid substrate and liquid, γ_{sl}, between substrate and gaseous atmosphere, e.g. vapor, γ_{sv}, and between liquid and atmosphere, γ_{lv}. The relation between these quantities and the contact angle Θ_Y of a droplet on the surface is described by Young's equation:

$$\cos \Theta_Y = \frac{(\gamma_{sv} - \gamma_{sl})}{\gamma_{lv}}. \tag{1}$$

The important factor is the energy loss following the increase in surface area of the droplet (i.e. spreading) relative to the energy gain following adsorption. The system reaches equilibrium when the total energy has reached a minimum. Detailed theoretical background can be found in [7, 8].

Besides the reduction of the substrate surface energy through the introduction of non-polar groups, the hydrophobicity can also be increased by increasing the surface roughness. Here, two potential cases have to be considered. If the liquid is able to penetrate the micro-rough structure, e.g. due to shallow surface profile and/or sufficient wetting, the contact area between liquid and substrate stays intact and is not interrupted by enclosed air. This is called the 'Wenzel state'. If the liquid, on the other hand, sits on top of the surface structure without penetrating the 'valleys', air will be enclosed between the droplet and the substrate and the effective surface of the droplet is increased. This is called the 'Cassie–Baxter state'. All discussed states are sketched in Fig. 1.

In the Wenzel (W) state the topographic properties of the surface are described by a roughness factor r [9], which gives the ratio of the effective area of actual, rough

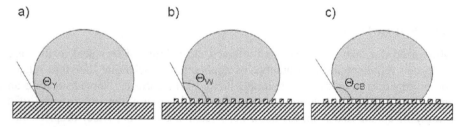

Figure 1. Behavior of a droplet on a perfectly flat surface (a), on a rough surface according to the 'Wenzel state' (b), and on a rough surface according to the 'Cassie–Baxter state' (c).

surface to the ideal flat surface, i.e. $r \geqslant 1$. The apparent contact angle is given by

$$\cos \Theta_W = r \cdot \cos \Theta_Y. \tag{2}$$

In the case of a flat surface r will be 1, accordingly $\Theta_W = \Theta_Y$. In the Cassie–Baxter (CB) state (cf. [2]) the liquid/air interface increases, while the solid/liquid interface approaches a minimum. The liquid can only gain very little adsorption energy and the increase of surface area (i.e. spreading of the droplet) is hindered for energetic reasons [7]. In the Cassie–Baxter state the apparent contact angle is given by

$$\cos \Theta_{CB} = -1 + \Phi_s(1 + \cos \Theta_Y), \tag{3}$$

where Φ_s is the ratio of liquid-solid interface area to the total apparent area ($\Phi_s \leqslant 1$). Again, given a perfectly flat surface, i.e. $\Phi_s = 1$, the contact angle is equal to the Young angle, $\Theta_{CB} = \Theta_Y$. One can easily take from Eqs. (2) and (3) that micro-roughening will in both states lead to increased hydrophobicity only if the ideal flat surface is already hydrophobic, i.e. if $\Theta_Y \geqslant 90°$.

The leaf of the Lotus plant has a surface topography with two scales of roughness in the form of a base profile with peak-to-peak distances of the order of several micrometers and a superposed fine structure with peak-to-peak distances significantly below one micrometer. Given this, the Lotus leaf follows the Cassie–Baxter state as sketched in Fig. 1c.

Surfaces with these characteristics are termed super-repellent because of a complete roll-off of a liquid droplet without any residues. The roll-off is accompanied by an effective cleaning of the leaf. Dirt particles are taken up by the droplet and removed. It has to be pointed out that such mechanism is valid for dirt particles stemming from the natural environment of these plants, e.g. sand or pollen. The validity for dirt compositions typical for industrial environments, which may have different behavior with regard to adhesion to the surface, adhesion to a rolling water droplet etc., cannot be assumed a priori.

As can be taken from the expressions given in Eqs. (2) and (3), a micro-roughening of a surface will effect an increase of the liquid repellence only if the surface is already hydrophobic (i.e. $\Theta \geqslant 90°$). In the case of a more hydrophilic surface ($\Theta < 90°$), however, the wettability will be increased. In the case of plants the 'intrinsic' hydrophobicity is due to cutin and epicuticular waxes. To adapt the principle to technical products, a suitable, hydrophobic finish may have to be combined with a treatment to create the micro-roughness.

Given the background of the self-cleaning property of the Lotus leaf, different approaches to create super-hydrophobic surfaces on textiles have been discussed (see e.g. [10, 11]). The deposition of nano-sized particles (fine structure) in a hydrophobic binder (see e.g. [12]) and the micro-roughening of fiber surfaces by means of UV laser irradiation [13] can be mentioned as examples. In the latter case, the irradiation of the polymer effects a change in the surface topography and generates a regular structure with a typical roughness between 2 and 3 μm [14–17]. Experimental results have shown that this treatment, especially in combination with a suitable hydrophobic finish, creates effects in agreement with the above discussion of Eqs. (2) and (3) [13]. Both stated cases make use of the general texture of the fabric which, such as in the case of fabrics made from multifilament yarn, represents a coarse base profile similar to plant surfaces. A rather recent paper by Gao and McCarthy [18] reports the exclusive use of the (existing) geometric structure of the fabric accompanied by a hydrophobic silicone finish. While the finish produces a contact angle of 110° on a flat surface, the authors report contact angles of up to 170° on a fabric made from poly(ethylene terephthalate) (PET) microfiber yarn. No further roughening of the fiber surfaces was incorporated.

Highly hydrophobic finishes are of increasing importance especially in the field of technical textiles. A large proportion of these products are finished with a heavy polymeric coating of, e.g., poly(vinyl chloride) (PVC), polyurethane, or silicone, which effectively masks the textile fabric with a smooth surface. One exemplary application of coated textiles, where easy or self-attained cleanability is highly required, is the construction of textile roofs in modern architecture. The present solution to decrease dirt take-up is to apply hydrophobic topcoats, i.e. layers of lacquer of approx. 5 μm thickness, to the coated fabrics. However, the performance of these conventional topcoats with regard to cleanability is not sufficient. As has been shown before, the approaches to create super-hydrophobic surfaces ([10–18]) refer to the texture of textile fabrics as well as the actual fiber surfaces but are not necessarily applicable to coated textiles.

A further important aspect is that the textiles considered here are subject to high, especially mechanical, stress. Textile roofs not only have to withstand weathering – UV aging, hail, etc. –, and high tensile forces, but also improper handling on the construction site. Micro-structured surfaces obtained by techniques such as, e.g., the described application of nano-sized particles [12] or laser treatment [13], but also by micro-lithography, are not likely to withstand these influences effectively.

Given this background, it was the objective of this work to study the potential of hydrophobic, UV cured topcoats with smooth as well as micro-rough surfaces to increase the hydrophobicity and cleanability of coated fabrics especially in urban and industrial environments. The recently introduced process of photochemically induced micro-folding (see Section 2) was employed to cure the topcoats. Since the process allows to generate rough as well as perfectly smooth surfaces of identical chemistry, special attention was paid to the relevance of the micro-rough topography to the stated surface properties. The samples were characterized with regard to topography, water contact angle, abrasion resistance, dirt take-up (in laboratory as well as in real-time aging), and cleanability.

2. THE PROCESS OF PHOTOCHEMICALLY INDUCED MIRCO-FOLDING

Although the fundamentals of the process of photochemically induced micro-folding are not the topic of this work, the background of this novel curing process is given in this section.

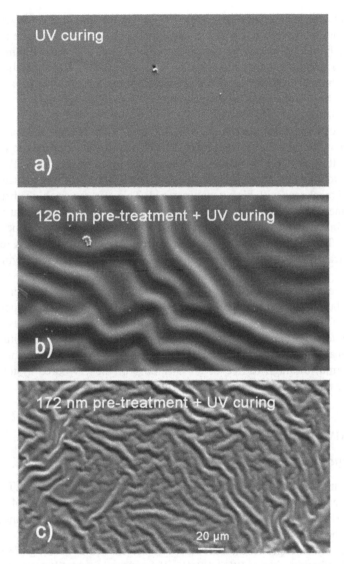

Figure 2. SEM micrographs of UV cured surfaces which had been produced from a lacquer coating of given chemistry (topcoat L2, cf. Table 2) with different, one-step (a) or two-step (b and c) curing procedures.

Recent papers published by the Leibniz-Institut für Oberflächenmodifizierung e.V. (IOM), Leipzig, Germany, report the structuring of UV curable coatings (lacquers) by photochemically induced micro-folding, effecting a micro-rough surface of this thin coating [19]. The extremely high absorption of UV wavelengths below

200 nm by acrylates results in a very fast curing of thin surface layer of about only 100 nm thickness on top of the still liquid bulk [20]. Due to the mechanical tensions (shrinkage) occurring during polymerization and cross-linking, a wavy surface topography is generated (Fig. 2). This structure is frozen-in through slow curing of the bulk volume, e.g. in a second step, by conventional broad-band UV irradiation from a medium pressure mercury lamp (250–320 nm), or by electron beam irradiation.

Commercial xenon excimer lamps emitting at 172 nm can be used for the energy deposition in the very thin surface layer, as well as novel windowless argon excimer lamps emitting at 126 nm, which have been developed at the IOM [19–20]. It is important to note that photons in the spectral range below 150 nm induce C-C- and C-H-bond breakages with high quantum yield. This effects very high absorbance in any polymerizable coating and, accordingly, extremely small penetration depths, resulting in intense folding with pronounced roughness and small peak-to-peak distances. The higher energy of photons with a shorter wavelength is responsible for significant differences in the surfaces resulting from treatments with 126 and 172 nm.

The process has been used mostly for matting/delustering of lacquer up to now [21–23], but basically it opens an interesting avenue to create super-repellent surfaces of coated fabrics, in accordance with the principles discussed. It is important to note that the relevant coating can be used as a substitute for the conventional topcoats often applied to these products. A further advantage can be expected from the high degree of cross-linking of the UV-cured systems, which might give rise to a high mechanical stability, e.g. against abrasion.

3. EXPERIMENTAL

3.1. Materials

Biaxially stretched film made from poly(ethylene terephthalate) (PET, thickness 100 μm) and industrial PET fabric finished with a white coating of poly(vinyl chloride) (PVC) served as substrates, to which the experimental topcoats were applied. The coated fabric was supplied by VS-Indutex GmbH, Krefeld, Germany. The water contact angle on the PET film was determined to be $\Theta = (73.0 \pm 0.9)°$, on the PVC coating as $\Theta = (86.3 \pm 2.4)°$. If employed for textile roofs, a protective topcoat is usually applied to the coated fabric. A corresponding commercial sample coated with a fluorinated lacquer was also supplied by VS-Indutex GmbH and used for comparison to the experimental samples. In this case, the contact angle was determined to be $\Theta = (86.0 \pm 1.0)°$.

The experimental lacquers under investigation were composed of a number of base compounds which are listed in Table 1. All systems were filled with SiO_2 nanoparticles in order to increase scratch resistance using Aerosil products by Degussa AG, Germany.

Table 1.
Components of experimental topcoats (lacquers)

Designation	Components
GL-A	Acrylated mono- and bi-functional monomers
GL-B	Modified vinylester resin + ethoxylated triacrylate
GL-C	polyfunctional (ethoxylated) aliphatic acrylates
GL-D	Acrylated bi-functional monomer, oligomeric tri- and hexa-functional urethane acrylates
GL-E	Aliphatic urethane acrylates, acrylated mono-, bi- and tri-functional monomers
GL-F	acrylated mono- and bi-functional monomers
GL-G	Aliphatic urethane acrylates, acrylated mono-, and tri-functional monomers
GL-H	Ethoxylated tri-acrylate, polyester acrylate
GL-J	Modified polyfunctional urethane acrylates
GL-C/E	Mix of GL-E and GL-C with additional bi- and tri-functional reactive thinners
AE1	Aerosil Ox50, Degussa AG, mean particle size 40 nm
AE2	Aerosil R972, Degussa AG, hydrophobized Aerosil, mean particle size 16 nm
AE3	Aerosil 380, Degussa AG, mean particle size 7 nm

3.2. Topcoat application

The topcoats were applied using labcoaters. UV curing was done under an inert atmosphere (nitrogen) with a dose of 670 $mJ \cdot cm^{-2}$ using a medium pressure mercury lamp with a specific power (electrical input) of 120 $W \cdot cm^{-1}$. The micro-folding, which was performed in the case of a number of systems before the normal curing step, was done with either a xenon excimer source emitting at 172 nm (Heraeus Noblelight, Hanau, Germany) or an argon excimer source emitting at 126 nm, which was developed and built at IOM.

3.3. Sample characterization

The surface topography of the top-coats was studied using a scanning electron microscope (SEM) (ABT 55, ISI, Korea) and characterized with regard to peak-to-peak distance S_y.

The surface properties were characterized by the (static) water contact angle Θ as a measure of wettability and the roll-off angle as a measure of "stick-slip" behavior of the droplet. The measurements were performed with the sessile drop method using a Krüss G40 analytical system (Krüss, Germany). Droplets of double-distilled water (10 μl) were placed on the sample surface at 20°C and the equilibrium contact angle determined. This angle was termed the static water contact angle. The data presented are mean values of 15 individual measurements. The roll-off angle Φ is the angle to which the substrate has to be tilted until the droplet rolls off.

The abrasion resistance of the (micro-structured) topcoats was characterized by measuring the contact angle Θ following severe mechanical treatments as well as by qualitative SEM analysis. The samples were scratched with a steel wool pad pressed against the sample with a defined load of 1 kg and rotated manually. The contact angle was measured after 10, 20 and 50 revolutions of the pad.

Dirt take-up was measured by spraying the samples with a standardized dirt solution and visual inspection. The dirt was a pigment and mineral oil mixture, supplied as "Fensterschmutz außen" by wfk-Testgewebe GmbH, Brüggen, Germany, which was dispersed in ethyl acetate. Before the application of the dirt, the samples were cleaned with water of high hardness (ion content 7.6 mmol/l) and left for one day.

Along with these laboratory tests, samples were also subjected to a 7-month real-time aging (roof aging) lasting from September to March. The samples were installed on the roof of building of the Deutsches Textilforschungszentrum Nord-West e. V. (DTNW) and tilted to 45° and oriented in southwest direction. This subjected the surfaces to rain, potential roll-off of water droplets and intense UV irradiation. Due to the geographic location, deposition of (oily) dust, pollen and gaseous emissions from industrial sources and traffic could be assumed.

The cleanability of the roof-soiled samples was characterized by means of a standardized cleaning procedure. A commercial wet cloth of size 3.5×4.5 cm^2 was applied with a pressure of 16 g/cm^2 and wiped over the sample once.

In all cases, the dirt take-up was characterized by deriving the degree of whiteness, W, from a colorimetric measurement. The dimensionless quantity can be calculated from the colorimetric data according to the Commission Internationale de l'Éclairage (CIE) standard [24, 25]. Given the white coating of the samples, taken-up dirt will effect a grey color shade and reduce the whiteness W. The measurement was performed with a Datacolor 3880 colorimetric system (Datacolor, Switzerland).

4. RESULTS AND DISCUSSION

4.1. Characterization of the experimental lacquers

Before the actual experiments on the generation of micro-structured surfaces by means of UV induced micro-folding, the experimental lacquers were studied with regard to scratch resistance, water contact angle Θ and roll-off angle Φ. For the following studies it was most important that the flat, unstructured coat had a good scratch resistance and a contact angle in excess of 90°. It should be pointed out again that an increase of hydrophobicity following micro-structuring of a surface can be expected only in the case of $\Theta_Y \geqslant 90°$ ('Young-angle' on a flat surface) according to the basic understanding of both Wenzel and Cassie–Baxter. The experimental lacquers which fulfill these criteria are listed in Table 2.

It is remarkable that in spite of notable hydrophobicity of the samples, high roll-off angles Φ were measured throughout, the only exceptions being samples 26 and 29. In the case of some samples no roll-off is seen at all. The observed "stick-slip" behavior is related to the contact angle hysteresis. In contrast to the static contact angle measurement – as was done in the framework of this study –, a dynamic measurement, e.g. according to the Wilhelmy method, allows to measure

Table 2.
Surface properties of the experimental topcoats (smooth surface) discussed in this work

Topcoat	Lacquer	Additives	Appl. weight [g/m^2]	Water contact angle Θ_Y [°]	Water roll-off angle Φ [°]
L1[a]	GL-C/E/AE2	UV stabilizer	5	108.3 ± 1.1	–
L2[a]	GL-C/E/AE2	UV stabilizer	5	107.7 ± 1.2	–
L3	GL-C/AE1	UV stabilizer, ethanol	5	110.2 ± 0.9	–
L4	GL-G (a)	UV stabilizer	5	106.9 ± 2.9	–
26	GL-E/AE2	1% hydrophobizing agent	6.8	99.8 ± 1.0	40 ± 5
27	GL-E/AE2	2% fluorinated acrylate	8.8	106.0 ± 1.2	>90
28	GL-E/AE2	1% hydrophobizing agent, 2% fluorinated acrylate	15	111.3 ± 0.5	81 ± 2
29	GL-D/AE2	1% hydrophobizing agent	7.9	98.9 ± 0.3	39 ± 7
30	GL-D/AE2	1% fluorinated acrylate	13.6	110.5 ± 1.4	>90
31	GL-D/AE2	1% hydrophobizing agent, 1% perfluorinated acrylate	6.6	102.8 ± 0.8	60 ± 17
32	GL-D/AE2	2% fluorinated acrylate	10.3	108.9 ± 1.4	–
40	GL-G 1	2% perfluorinated acrylate	12	112.4 ± 0.4	–
41	GL-G 1	2% perfluorinated acrylate, 0.5% stabilizer, 1% UV absorber	6.8–13.7	108.6 ± 0.7	–
42	GL-G 1	1% hydrophobizing agent, 2% perfluorinated acrylate, 0.5% stabilizer, 1% UV absorber	6.3–10.4	105.0 ± 0.4	–
43	GL-G 1	1% hydrophobizing agent, 0.5% stabilizer, 1% UV absorber	8.4–10.1	99.5 ± 0.2	–
44	GL-H	20% corundum	10.1	98.8 ± 0.8	–

[a] Topcoats L1 and L2 are composed of identical components, but were prepared with slightly differing formulations.

the advancing angle Θ_A and the receding angle Θ_R. The difference of these two angles is called the contact angle hysteresis $\Delta\Theta$ ($\Delta\Theta = \Theta_A - \Theta_R$). In the case of a high hysteresis, a droplet will have a high roll-off angle in spite of a hydrophobic surface. Super-hydrophobic surfaces will have a hysteresis $\Delta\Theta$ close to 0.

4.2. Properties of micro-structured surfaces

On the basis of the above results, a number of lacquers were applied to PET film and cured following different procedures in order to generate smooth – i.e. perfectly flat – as well as rough surfaces. Accordingly, the samples were either conventionally cured or following the described process of micro-folding, i.e. cross-linking only at the surface in a first step (pre-curing), followed by bulk curing in a second step. The surface curing was done using either a 126 nm source (argon excimer) or a 172 nm source (xenon excimer).

Table 3.
Static water contact angles on lacquer topcoats with smooth (one-step cured) as well as micro-structured (two-step cured) surface. The topcoats L1, L2 and L4 were applied onto PET film, and topcoats 26 and 29 onto industrially PVC-coated PET fabrics

Topcoat	Contact angle Θ [°] following		
	(conv.) UV-curing	126 nm + UV-curing	172 nm + UV-curing
L1	111.1 ± 2.2	109.8 ± 1.7	106.9 ± 5.9
L2	109	111.6	107.3 ± 0.6
L4	108.5	103.9	109.8
26	99.8 ± 1.0	–	96.1 ± 0.8
29	98.9 ± 1.2	–	95.0 ± 1.3

Exemplary SEM micrographs of samples treated in the described way are shown in Fig. 2. It is notable that coatings of identical chemistry can be designed to have different surface topographies depending on the curing procedure and the absorption properties of the lacquer for the chosen wavelength(s). It is also worth noting that certain lacquers did not show any micro-folding no matter how the curing was done.

From SEM micrographs, the characteristic parameters of the surface profile can be estimated. It can be derived from the analysis that the micro-folded samples typically have surface topographies with peak-to-peak distances S_y of 10 to 20 μm.

A central aspect of this part of the work was to study the (potential) effect of the surface structure generated by micro-folding on the wetting behavior of a water droplet. Accordingly, water contact angles were measured on the topcoat systems, which could be photochemically micro-folded. Table 3 shows the measured values on the flat (one-step cured) as well as the micro-rough (two-step cured) surfaces. The data clearly show that an increase of the contact angle was not found in any case, Θ being approximately constant within experimental accuracy, hence independent of the curing procedure. As has been discussed in Section 1, one has to consider the so-called aspect ratio of the surface profile, i.e. height vs. width of the 'grooves' in order to differentiate the different cases – Young, Wenzel, Cassie–Baxter – of how a droplet wets the surface. The height is given by the root mean square (rms) roughness R_{rms} and the width of the 'grooves' by S_y. The aspect ratio is influential on the effective contact area of a droplet and also on the probability that the droplet (liquid) penetrates into the surface structure (transition from Wenzel to Cassie–Baxter state). Even if the SEM micrographs do not give any quantitative measures, small aspect ratios are indicated throughout. In an earlier study the authors characterized the topography of micro-folded lacquer layers using the same UV curing process by means of white-light interferometry and found aspect ratios of the order of 0.1 [26]. While the lacquer layers in question were of different chemistries and were applied onto different substrates, the value may give an indication of the topography of the samples studied here. From the resulting small aspect ratio, a value of Wenzel's roughness factor r (Eq. (2)) of the order of

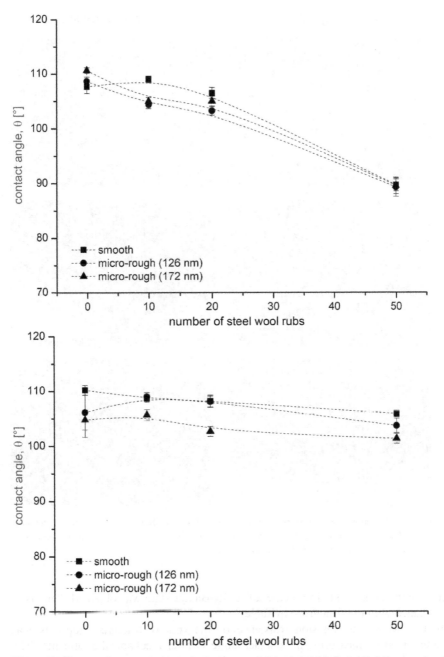

Figure 3. Change of water contact angle on topcoats L2 (top) and L3 (bottom) following abrasion (rubbing with a steel wool pad). Note: The topcoat L3 does not show the effect of micro-folding.

original   after abrasion

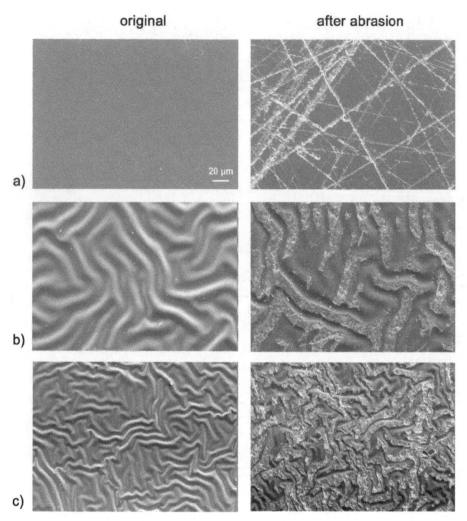

a)

b)

c)

Figure 4. SEM micrographs of topcoat L2 on PET film before and after abrasion (50 rubs with a steel wool pad). The curing was done either conventionally (a) or in the described two-step process using either 126 nm (b) or 172 nm UV (c) for the pre-curing.

unity can be concluded. In no case a Cassie–Baxter state was obtained. This is in accordance with the data given in Table 3.

In this work, the abrasion resistance of the (micro-structured) topcoats was characterized by measuring the contact angle Θ after mechanical treatments. The graphs shown in Fig. 3 give exemplary results for two different systems. Depending on the basic hardness of the lacquer, a clear decrease of the water contact angle is found. As a consequence of the surface damage, the contact angle falls below

90° (cf. Fig. 3, top graph). It is to be noted that the phenomenon is also found, to some extent, in the case of systems which were only bulk cured – i.e. have a flat, unstructured surface –, as well as in the case of the systems which do not show any micro-folding at all.

Accordingly, one has to assume a damage not to the micro-rough surface, but to the bulk of the topcoat. This is also indicated by the SEM micrographs shown in Fig. 4.

4.3. Dirt take-up and cleanability

The observations discussed indicate that the micro-structuring of experimental lacquers (topcoats) by the process of UV induced micro-folding does not lead to the envisaged increase in hydrophobicity. In no case, the Cassie–Baxter state of a roughness-dependent contact angle was achieved. One has to conclude that the surface topographies do not yield the correlation between super-repellence and self-cleaning as is known from the Lotus leaf. It was pointed out before, however, that the mechanism is valid for the dust prevailing in the natural environment of the plant, e.g. sand or pollen. One cannot assume a priori that this is also valid for the dirt as found in industrial areas with potentially different adhesion to surfaces, take-up by a water droplet, etc.

Given this background, the dirt take-up behavior of the experimental topcoats was studied extensively. Assuming typical fields of applications of technical outdoor products – e.g. roofs, tarpaulins, blinds, etc. –, the measurements here concentrated on dirt typical for industrial and urban areas. The pure PVC coating and a ready-made fabric finished with a conventional protective topcoat were considered for comparison.

In order to characterize dirt take-up, the samples were sprayed with a standardized pigment- and oil-containing dirt and their degree of whiteness measured. In comparison to conventional systems – e.g. fluorinated lacquers –, all experimental topcoats had a reduced take-up of oil- and pigment-containing dirt typical for urban and industrial environments. The photograph shown in Fig. 5 gives an exemplary result from these experiments. The sample exhibits a heavy dirt deposition only on its edges, which carries no coating. The marked difference in dirt take-up of the coated area is clearly evident.

For a quantitative assessment, the degree of dirt take-up was determined by colorimetry and calculation of the whiteness W according to the CIE-system and, especially, the change in whiteness ΔW. Given the white color of the underlying PVC coating, the whiteness was expected to give a sensitive measure of the amount of dirt deposited on the samples. The evaluation of the data showed that the samples had significantly different degrees of dirt take-up, as shown in Fig. 6. There is no correlation between the change in whiteness and the water contact angle on the samples, however.

On the basis of these results, a set of 10 samples was subjected to 7 months of real-time aging. The experiment lasted from September to March and was

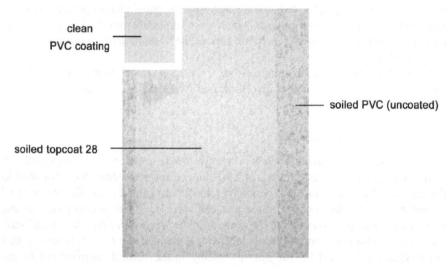

Figure 5. Photographic view of a sample which had been soiled, i.e. sprayed with standardized dirt, under defined conditions in the laboratory.

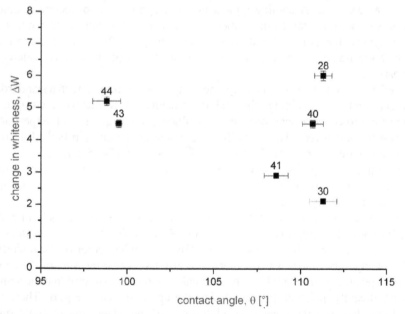

Figure 6. Change in whiteness ΔW following the defined application of dirt in the laboratory vs. water contact angle on the (clean) samples. Numbers denote topcoats according to Table 2.

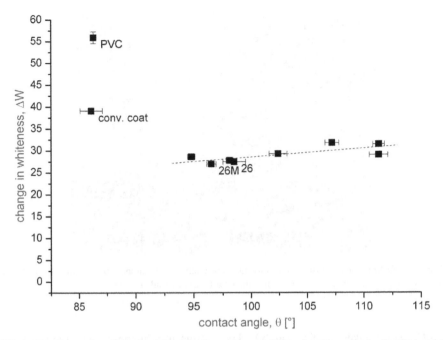

Figure 7. Change in whiteness ΔW after 7 months real-time aging in a mixed urban/industrial environment vs. water contact angle on the (clean) samples. Numbers denote topcoats according to Table 2. The topcoats 26 and 26M are systems of identical chemistry but with smooth and micro-rough surfaces, respectively.

meant to reflect the sum effect of simultaneous processes, namely dirt take-up from particulate and gaseous emissions, cleaning (by rain), and surface aging or damage through weathering and UV light.

Based on experimentally determined contact angle and abrasion resistance, the set included samples of identical chemistry with both smooth and micro-rough surfaces. Again, the pure PVC coating and a ready-made fabric finished with a conventional protective topcoat were considered for comparison. The samples were installed on the roof of the DTNW and tilted to 45° and oriented in southwest direction. This subjected the surfaces to rain, potential roll-off of water droplets and intense UV irradiation and simulated the typical application conditions of technical outdoor products.

The characterization of the samples by colorimetry showed marked differences in the whiteness W after 7 months aging. For the systematic evaluation of the data, the change in whiteness ΔW following the test was calculated and correlated to the water contact angle (cf. Table 2). This is shown graphically in Fig. 7. It is evident that all experimental topcoats show a superior dirt take-up behavior, i.e. less ΔW, compared to the conventional system (despite similar wettability, the pure

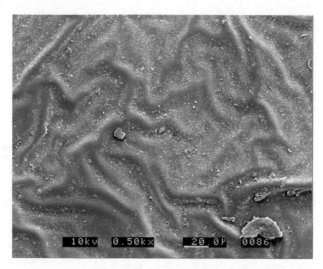

Figure 8. SEM micrograph of sample 26M (identical chemistry as sample 26, but with a micro-rough surface) after 7 months real-time aging in a mixed urban/industrial environment.

PVC coating takes up dirt heavily). The change in whiteness – i.e. the dirt take-up – is low in the case of the hydrophobic topcoats, but does not decrease further with increasing contact angle. The comparison of the data on lacquer system 26 with smooth and micro-folded surface (denoted 26M in Fig. 7) does not indicate a positive effect of the micro-rough surface on the extent of dirt take-up.

SEM micrographs of the samples taken before and after the real-time aging give a qualitative impression of the different deposition of particulate matter on experimental topcoats with smooth surfaces compared to the industrial product which was finished with a conventional, fluorinated topcoat. The majority of the dirt/dust particles are smaller than 2 μm. The SEM analysis of a micro-structured surface revealed that the rough profile 'embeds' especially these small particles (Fig. 8). This might indicate a detrimental effect of the micro-structuring with regard to cleaning the surface from well-adhering dirt particles.

Given this background, the samples that had been subjected to real-time aging were cleaned in a simple, well-defined manner and again characterized with regard to their whiteness. To remove the deposited dirt, the samples were wiped with a commercial wet cloth with constant pressure and wiping procedure. The corresponding data are given in Fig. 9, which includes also the already shown data taken after aging, but before cleaning (from Fig. 7). It can be taken from measurement of the whiteness that the dirt load can be removed very effectively – at times totally – in the case of the experimental topcoats. In all cases, the cleaning was more efficient than with the conventional sample. The PVC coating – i.e. without any topcoat – cannot be cleaned from the deposited dirt at all. It is important to

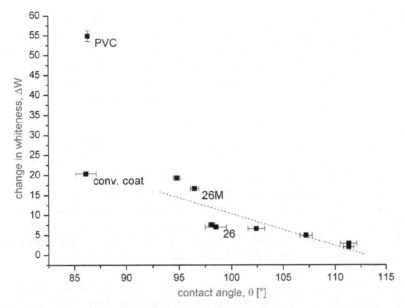

Figure 9. Change in whiteness of various samples following real-time aging and ensuing cleaning in dependence on the water contact angle. Numbers denote topcoats according to Table 2. The topcoats 26 and 26M are systems of identical chemistry but with smooth and micro-rough surfaces, respectively.

note that a sample with micro-structured surface cannot be cleaned as efficiently as the corresponding smooth sample of identical chemistry, which is in accordance with the observation of 'embedded' particles made in the SEM analyses discussed before.

5. SUMMARY

The effect of the water wettability of lacquer topcoats on coated textiles on their dust take-up and cleanability was studied. In accordance with present understanding of the self-cleaning mechanisms of plant and animal surfaces, the topcoats were designed to be highly hydrophobic. Besides the chemical composition of the lacquer topcoats, topographic effects were considered, namely the generation of a micro-roughness with the help of a novel UV curing process. In this process UV irradiation with a wavelength of 126 or 172 nm is used for the cross-linking of an ultra-thin surface layer, while the bulk of the lacquer is cured subsequently at longer wavelengths. Following shrinkage a rough surface is obtained ('micro-folding').

The experimental lacquers (smooth surfaces) had water contact angles on the order of 100 to 115° and rather good scratch resistance. In none of the studied systems, however, did the micro-structuring of the surfaces by UV induced micro-folding lead to the envisaged increase in hydrophobicity. SEM analyses revealed that the surfaces of samples pre-cured with 126 or 172 nm typically had topographies with peak-to-peak distances S_y of 10 to 20 μm. Even if the aspect ratios of the surface profiles could not be determined quantitatively from the SEM micrographs, small values were indicated in all investigated cases. It can be concluded that in no case a Cassie–Baxter state of a roughness-dependent contact angle was achieved.

The dirt take-up behavior of the experimental topcoats was studied extensively. Assuming typical fields of applications of outdoor products – e.g. roofs, tarpaulins, blinds, etc. –, the measurements concentrated on dirt typical for industrial and urban areas. In comparison to conventional systems – e.g. fluorinated lacquers –, all experimental topcoats had a reduced take-up of oil- and pigment-containing dirt typical for urban and industrial environments. No correlation was found between the change in whiteness and the water contact angle on the samples, however. It can be assumed from these findings that the mechanisms responsible for the self-cleaning of certain plant and animal surfaces is not necessarily valid for dirt as found in industrial areas with potentially different adhesion to surfaces, take-up by a water droplet, etc.

Thus, real-time aging experiments over 7 months were performed and revealed a reduced dirt take-up by all experimental topcoats compared to the conventional system. The dirt take-up was low in the case of the hydrophobic topcoats, but did not decrease further with increasing contact angle. The data did not indicate a positive effect of the micro-rough surface on the extent of dirt take-up.

Acknowledgements

The authors wish to thank the Forschungskuratorium Textil e.V. for their financial support for this project (AiF-Nr. 13874 BG). This support is granted from resources of the Bundesministerium für Wirtschaft und Technologie (BMWi) via a supplementary contribution by the Arbeitsgemeinschaft Industrieller Forschungsvereinigungen "Otto-von-Guericke" e.V. (AiF).

REFERENCES

1. G. E. Fogg, *Nature* **154**, 515 (1944).
2. A. B. D. Cassie and S. Baxter, *Nature* **155**, 21-22 (1945).
3. A. A. Abramzon, *Khimia i Zhizn* **11**, 38-40 (1982).
4. S. Shibuichi, T. Onda, N. Satoh and K. Tsujii, *J. Phys. Chem.* **100**, 19512 (1996).
5. K. Tsujii, T. Yamamoto, T. Onda and S. Shibuichi, *Angew. Chem.* (Engl. edition) **109**, 1042 (1997).
6. W. Barthlott and C. Neinhuis, *Planta* **202**, 1 (1997).
7. W. Nachtigall, *Bionik – Grundlagen und Beispiele für Ingenieure und Wissenschaftler*, Springer, Berlin (1998).
8. K. L. Mittal (Ed.), *Contact Angle, Wettability and Adhesion*, Vol 4, VSP/Brill, Leiden (2006).
9. R. N. Wenzel, *Ind. Eng. Chem.* **28**, 988 (1936).
10. W. Barthlott and C. Neinhuis, *Intl. Textile Bulletin* **47**, 8-12 (2001).

11. M. C. Thiry, *AATCC Review* **4**, 9-13 (2004).
12. T. Stegmaier, M. Dauner, A. Dinkelmann. A. Scherrieble, V. von Arnim, P. Schneider and H. Planck, *Technical Textiles* **47**, 186-191 (2004).
13. T. Bahners, E. Schollmeyer and D. Praschak, *Melliand Textilber.* **82**, 613-614 (2001).
14. T. Bahners, D. Knittel, F. Hillenkamp, U. Bahr, C. Benndorf and E. Schollmeyer, *J. Appl. Phys.* **68**, 1854 (1990).
15. T. Bahners, W. Kesting and E. Schollmeyer, *Appl. Surface Sci.* **69**, 12-15 (1993).
16. T. Bahners, *Optical Quantum Electronics* **27**, 1337-1348 (1995).
17. T. Bahners, T. Textor and E. Schollmeyer, in: *Polymer Surface Modification: Relevance to Adhesion*, Vol. 3, K. L. Mittal (Ed.) pp. 97-123, VSP/Brill, Leiden (2004).
18. L. Gao and T. J. McCarthy, *Langmuir* **22**, 5998-6000 (2006).
19. C. Elsner, M. Lenk, L. Prager and R. Mehnert, *Appl. Surface Sci.* **252**, 3616 (2006).
20. European Patent EP 1050395 B1 (2006).
21. German Patent DE 4439350 C2 (1997).
22. US patent 005888617 (1999).
23. German patent DE 19842510 A1 (2000).
24. R. Hützeler, *Textilveredlung* **18**, 151-156 (1983).
25. R. Griessner, *Textilveredlung* **18**, 157-161 (1983).
26. T. Bahners, K. Opwis, E. Schollmeyer, B. Marquardt and L. Prager, *Selbstreinigende Lack-Topcoats auf Planen und Membranen durch photonisch induzierte Mikrofaltung*, DTNW-Mitteilung issue no. 53 (2006), ISSN 1430-1954.

Contact Angle, Wettability and Adhesion, Vol. 5, pp. 173–190
Ed. K.L. Mittal
© VSP 2008

Wetting and adhesion in fibrous materials:
Stochastic modeling and simulation

WEN ZHONG,[1,*] NING PAN[2] and DAVID LUKAS[3]

[1]*Department of Textile Sciences, University of Manitoba, Winnipeg, MB R3T 2N2, Canada*
[2]*Department of Biological and Agricultural Engineering, University of California, Davis, CA 95616, USA*
[3]*Technical University of Liberec, Liberec, Czech Republic*

Abstract—Wetting and adhesion occur in the processing and use of fibrous materials when an interface is created between the fibrous substrate and a liquid or solid (e.g., coating or resin). Fibrous materials have a unique structure of complex geometry, with pores distributing from intra-fiber to inter-fiber spaces. But, physically speaking, wetting and adhesion are due to molecular interactions within a solid or liquid or across the interface between a liquid and a solid. Stochastic modeling and simulation methods (e.g. The Ising model combined with the Monte Carlo simulation) are, therefore, used in the studies of wetting and adhesion behaviors of fibrous materials, including liquid wetting/wicking in fibrous structures, single fiber pull-out process in a composite, and tearing behavior of coated fabrics. Complicated mechanisms of these behaviors can be realistically simulated with a relatively simple algorithm. Namely, the interaction across the interface (cohesion within a single substrate, adhesion between different meterials) can be represented by a Hamiltonian expression for the system, so that minimization of the system Hamiltonians will yield the most likely new steps for system evolvement, while the Monte Carlo method can be used to select the step that will actually occur, reflecting the stochastic agreement with the behavior of real systems. This approach is demonstrated to be useful in the studies of interfacial phenomena.

Keywords: Wetting; adhesion; fibrous materials; simulation.

1. INTRODUCTION

As frequently observed phenomena in the processing and use of fibrous materials, wetting and adhesion have been the object of much research. For instance, liquid transport in fibrous materials has attracted continued research interest in the fields of fiber processing and fiber composite manufacturing, especially when efforts are made in such areas as production of nano-fibrous materials for absorption, and bio-filtration and purification, which increasingly demand more precise and efficient methods for studying this phenomenon.

---

*To whom correspondence should be addressed. Tel.: 1-204-4749913; Fax: 1-204-4747593; e-mail: Zhong@cc.umanitoba.ca

Much literature is available concerning the wetting process on a solid surface. This includes several comprehensive reviews [1, 2], covering topics from contact angle, contact line, liquid/solid adhesion, to wetting transition and dynamics of spreading. Although directed towards the same goal, our work differs from previous researches because we treat wetting processes that are complicated by the interaction of a liquid with a porous medium of *intricate*, *tortuous* and yet *soft* surface, instead of a simple solid one.

Most textile processes are time limited, so the kinetics of wicking becomes very important. The classical Washburn equation [3] describes the liquid velocity moving up or down in a perpendicular capillary, neglecting the inertia of the liquid column. To apply the Washburn equation to wicking studies, Minor [4, 5] used several kinds of liquids in his wicking experiments and established that the wicking height of the liquid in a fiber or yarn was proportional to the square root of the time. Actually, however, the liquid column will cease to rise after a certain period of time due to the balance of surface tension and gravity. Accordingly some work has been dedicated to the modification of the Washburn equation [6].

Wicking is also affected by the morphology of fibrous assemblies. Most fibrous structures are never of a perfect capillary [7, 8], and can, therefore, further complicate the problem [9]. To tackle such problems, some researchers tried to derive equivalent capillary radius in their studies [10]. They did not, however, report efficient, precise ways of representing the intricate structure of fibrous materials.

Also, wetting of fibrous materials is dramatically different from the wetting process on a flat surface, due to the geometry of the cylindrical shape. A liquid that fully wets a material in the form of a smooth planar surface may not wet the same material when it is in the form of a smooth fiber surface. Brochard [11] discussed the spreading of liquids on thin cylinders, and stated that for nonvolatile liquids, a liquid drop cannot spread out over a cylinder if the spreading coefficient S is smaller than a critical value S_c, instead of 0. Also plenty of research work has been published on the equilibrium shapes of liquid drops on fibers [12–16]. Two distinctly different geometric shapes of droplet are reported: a barrel and a clamshell.

As a precise description of the structure of a fibrous material can be tedious, many researchers adopted Darcy's law, an empirical formula that describes laminar and steady flow through a porous medium in terms of the pressure gradient and the intrinsic permeability of the medium [17–19]. Darcy's law reflects the relationship of pressure gradient and average velocity only on a macroscopic scale, and it cannot describe the microscopic details of the liquid penetrating fibrous media.

In the 1990s, Manna et al. [20] presented a 2D stochastic simulation of the shape of a liquid drop on a wall due to gravity. The simulation was based on the so-called Ising model and Kawasaki dynamics. Lukkarinen et al. [21] studied the mechanisms of liquid droplet spreading on flat solids using a similar model. However, their studies dealt only with flow problems on a flat surface instead of a real heterogeneous structure. Only recently the Ising model has been used in the simulation of wetting dynamics in heterogeneous fibrous structures [22–26].

On the other hand, adhesion between two solids usually involves fiber reinforced composites or coated fabrics. Study of the mechanisms of the matrix-fiber interfacial interactions is now one of the key issues in composite studies.

Experimental techniques were developed to characterize the interfacial properties in a fiber reinforced composite [27]. Among them, the pull-out test seems to be the most direct approach to studying composite interfacial behavior. In a pull-out test, a single fiber embedded in a block of matrix (cured resin) is pulled out by a tensile load. The process of having the fiber pulled out from the matrix is observed to demonstrate several typical stages: fiber debonding, post-debonding friction of fiber against debonded surfaces, and, finally, either the fiber breaking or matrix being pulled out by the fiber. The simple pull-out test and its theoretical concept has been further developed to study mechanical behavior of a more sophisticated composite material [28].

To better interpret the experimental results, numerous theoretical models have been developed for describing the pull-out test, but theoretical analysis of this seemingly simple test method remains to be a challenge. Generally, these analyses were classified into two categories [29], as follows.

In the approach based on shear strength criterion, debonding occurs when the interface shear stress reaches the interface shear bond strength. The shear stress distribution along the interface is usually evaluated in this approach [30, 31]. In spite of the tremendous efforts to validate these theoretical models, a reasonably satisfactory closed analytical solution is still not available. Although the Finite Element Method (FEM) provides a more precise tool to solve the models numerically, it is difficult to incorporate such non-linear factors as the interfacial friction and matrix inelasticity in the FEM analysis, mainly because a non-linear analysis requires a dramatically increased computational expenditure [32, 33]. In the energy-based approach, extension of a crack requires the potential energy release rate of the composite constituents to reach a critical value, i.e., the interface fracture toughness [34, 35]. The energy-based models of the pull-out process can also be solved with the aid of FEM [36].

When comparisons are made between the strength-based and the energy-based approaches, it is generally agreed that the strength theory neglecting friction can be applied to a totally unstable debonding process with short embedded fiber lengths, whereas the energy method can only describe stable debonding usually with long embedded fiber lengths [35, 37]. But an acceptable model for describing the more frequent behavior of partial debonding is not available yet [37].

Different approaches are used to deal with other interface problems involving wetting and/or adhesion. Actually, these seemingly different topics or research areas (i.e., the problem of adhesion at the interface between two solids and liquid wetting/transport phenomena) share common basis. It is, therefore, quite possible that they can be studied by the same or similar approach.

This paper is about the use of a stochastic model (i.e., the Ising model combined with the Monte Carlo simulation) and its application in the investigation of both the wetting process of a liquid on fibrous substrates and adhesion/debonding at the interface between two solid phases in a composite material. With this approach, the statistical genesis of the process of liquid penetration in fibrous media (or adhesion/debonding between fiber and resin) can be regarded as the interactions and the resulting balance among the fiber and liquid (or resin) cells that comprise the ensemble. These processes are driven by the difference in internal energy of the system after and before liquid movement from one cell to the other (or partial debonding, fiber breaking, or matrix failure at a fiber or resin cell). This approach can be adapted to study other interface problems involving wetting or adhesion, where there is a system that consists of interactive subsystems, each with two interchangeable states, i.e., wetting or nonwetting, adhesion or debonding.

2. LIQUID WETTING/WICKING IN FIBROUS STRUCTURES

2.1. Model description

In statistical thermodynamics, macro-characteristics of a system are due to the interactions and resulting balance among micro-particles or cells that comprise the system. In the original Ising model, accordingly, a one-dimensional system is divided into a number of lattice cells. Then the Hamiltonian is calculated as the summation of the interactions between each pair of the nearest neighboring cells. And the average macroscopic parameters of the system can be derived from the Hamiltonian via the route of statistical mechanics. The advantage of the Ising model is that, due to its simple expression, it can be used to describe a complex system made of subsystems with two interchangeable configurations, i.e., by "digitalizing" the original system into a grid of cells with state 1 or 0 only.

In a two-dimensional Ising model for liquid transport, the field is divided into a lattice frame of $L \times L$ square cells, as shown in Fig. 1. Two variables are used in the model to describe the state of each cell in the lattice:

i) s is used to describe whether a cell is occupied by the liquid denoted as 1, or 0 otherwise. In the simulation, liquid can move from one cell to another, indicating transport of liquid in fibrous materials.

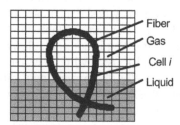

Figure 1. Two-dimensional Ising model.

ii) F is used to describe whether a cell is occupied by the fibrous medium, taking either 1 (a cell with fibers) or 0 (a cell without fiber). For the sake of simplicity, fibers are assumed to be free from moving during the simulation.

The total Hamiltonian of the system, H, is considered as a sum of all the contributions of the cell energies in the lattice. When interactions between the nearest neighbors only are considered, the Hamiltonian for an arbitrary cell i can be expressed as

$$H_i = - \left(B_0 \sum_i s_i F_i + B_1 \sum_{i,j} s_i F_j \right) - C \sum_{i,j} s_i s_j + G \sum_i s_i y \qquad (1)$$

where the first term in the bracket represents the interaction between liquid and fiber substrate spins that coexist in one cell while the second term represents those in the neighboring cells, the coefficient B_0 is the adhesion energy of interaction between fiber and liquid within the cell i, and B_1 is the adhesion energy of fiber/liquid interactions between cell i and its neighboring cells. The second term represents the interaction between liquid cells, and C is the cohesive energy of liquid. The last term represents the energy of the liquid spin in the gravitational field, of which G is the intensity of the gravitation field and y is the y-coordinate of the cell in the lattice. Interaction of more distant terms can be incorporated when necessary [22].

2.2. Simulation algorithm

For the sake of simplicity, in all the simulations discussed in this study, swelling/movement of fibers is neglected, i.e., the values of F for all the cells are kept constant in the simulation. The process of a liquid penetrating a fibrous structure is, therefore, the result of the liquid occupying empty spaces between fibers and wetting the fibers. The various steps of the simulation are as follows:

(i) An initial configuration is created by developing the lattice, on which the fibrous medium is placed for further simulation. The initial values of both F (1 or 0) and s (1 or 0) for each cell are determined. A cell i is considered to be covered with fiber ($F = 1$) if the distance between the cell center and the fiber axis is smaller than the fiber radius. And the number of cells near the injection port is so defined that their s values are always $+1$, indicating that there is a constant influx of liquid.

(ii) The liquid-fiber interface is scanned. A cell i in the lattice is randomly selected. And, if the values of s for one or more nearest neighbor cells are different from cell i, cell i is supposed to be on the interface. As shown in Fig. 1, the dry cell i is next to three liquid cells and five liquidless ones, each liquid cell having the opportunity to exchange spin values with cell i, but with different probabilities.

(iii) Cell i can be paired with each of the three liquid cells, respectively. For each pair, ΔE_T, which is the energy difference between the configurations before and after the exchange of spin values of the two cells, can be calculated. Then the pair with the lowest value of ΔE_T is selected as the most probable exchange. If a random number uniformly distributed between 0 and 1 is less than the spin-flip probability, $p = \exp(-\beta_\Delta E_T)$, the exchange of two spins takes place. Here, β is a constant inversely proportional to the absolute temperature.

(iv) A Monte Carlo Step (MCS) ends when all the cells at current liquid-fiber interface have been scanned. Then step (ii) is repeated to start a new MCS until the simulation is terminated.

2.3. Simulation results and discussion

The coefficients, B_0, B_1 and C in Equation (1), which are essential parameters to the Ising model for transport, are determined in the following manner: As ubiquitous interactions across interfaces, the van de Waals forces can be assumed to dominate the interactions between fiber and liquid. Other interactions, like electrostatic forces, acid/base interactions, could be significant in some materials/liquid interactions. However, they will not be considered here. According to the Lifshitz theory [38], the interaction energy per unit area between two surfaces can be expressed as

$$W_{1,2} = \frac{-h_{1,2}}{12\pi D^2} \qquad (2)$$

where $h_{1,2}$ is the Hamaker constant and D is the distance between the surfaces. Derived from the quantum field theory, an approximate expression for the Hamaker constant of two bodies (1 and 2) interacting across a medium 3, none of them being a conductor, is

$$h_{1,2} = \frac{3h v_e (n_1^2 - n_3^2)(n_2^2 - n_3^2)}{8\sqrt{2}(n_1^2 + n_3^2)^{1/2}(n_2^2 + n_3^2)^{1/2}\{(n_1^2 + n_3^2)^{1/2} + (n_2^2 + n_3^2)^{1/2}\}}$$
$$+ \frac{3}{4}k_B T \frac{\varepsilon_1 - \varepsilon_3}{\varepsilon_1 + \varepsilon_3} \cdot \frac{\varepsilon_2 - \varepsilon_3}{\varepsilon_2 + \varepsilon_3} \qquad (3)$$

where h is the Planck's constant, v_e is the main electronic absorption frequency in the UV (assumed to be the same for the three bodies, and typically around 3×10^{15} s^{-1}), and n_i is the refractive index of phase i, ε_i is the static dielectric constant of phase i, k_B is the Boltzmann constant, and T the absolute temperature.

The constants k_1 and k_2 are used to represent the ratios of B_1/C and B_0/B_1, respectively.

$$k_1 = \frac{B_1}{C} = \frac{W_{1,2}}{W_{1,1}} = \frac{h_{1,2}}{h_{1,1}}, \qquad k_2 = \frac{B_0}{B_1} = \frac{W_{1,2}\pi dal}{W_{1,2}a^2} = \frac{\pi dl}{a} \qquad (4)$$

The ratio B_0/B_1 is equal to the ratio of adhesive surface areas within a cell and that between neighboring cells. The area for coefficient B_0 is calculated as the total surface of l individual filaments (cylindrical with diameter d) in the yarn within a

Table 1.
Specifications of PP yarn samples used

Sample	Yarn count (tex)	Diameter of yarn (mm)	Filament fineness (dtex)	No. of filaments in a yarn	Filament diameter (μm)	Total surface area of fiber in a single cell (mm^2)
1	13.4	0.282	2.25	60	18	0.956
2	13.4	0.290	1.5	90	14.6	1.197
3	13.4	0.322	0.6	224	9.2	2.084

cubic cell (with side length a), assuming that filaments within a yarn are evenly packed and stretched straight along the axis of the yarn, and the area for coefficient B_1 is calculated as the area of the cell. The value of B_1 is determined by simulation to accommodate the experimental data, and the values of C and B_0 are determined by Equation (4).

To test the validity of the model, a set of wicking experiments were performed. Test samples included three types of weakly twisted polypropylene (PP) filament yarns with the same yarn count but different fineness of constituent filaments. Their specifications are listed in Table 1. Each sample was hung vertically by a clamp and the free end was dipped into a beaker containing 1% solution of Methylene Blue dye. Traveling height of the liquid was measured during the wicking tests. To obtain the wetting rate, the time required for the dye solution to travel upwards along the sample was recorded.

In the simulation, the plane is divided into 9 × 150 square cells. In order to conveniently calculate the weight of the liquid in a cell and the interfacial area between two cells, each cell is supposed to be a cubic. The wicking experiments were performed on weakly twisted polypropylene filament yarns. Weak twists were used to maintain the round shape and evenness of yarn cross sections along the lengthwise direction, so that a filament yarn can be represented as a fibrous assembly with even diameter: the width of a cell is made equal to the diameter of the yarn in each case. On the other hand, the weak twists somewhat pack the fiber assemblies to provide a large number of capillaries (interstices between filaments) in the lengthwise direction as wicking channels. The weak twists also enable the simulation to depict the fiber layout as parallel to yarn axis in each cell.

Polypropylene fiber, a fiber material with zero moisture absorbency, was chosen to eliminate any shape change (i.e. swelling) of individual filaments during the experiment. As a result, the surface area of fibers is assumed to be constant during the simulation. The dominant factor for the wicking tests, therefore, is the structural differences between the filament yarns with the same yarn count but different diameters for constituent filaments.

The seed of the random number is set to be 0 at the beginning of each simulation to make the simulation repeatable. Let each Monte Carlo step last for 1 second. Parameters needed for the simulation are listed in Table 2. The experimental as well as the simulation results of the wetting rate are shown in Fig. 2. The constant

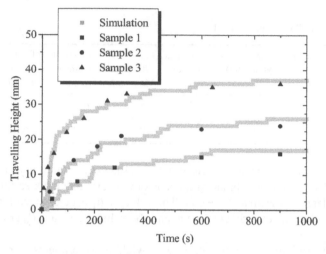

Figure 2. Traveling height of liquid versus time.

Table 2.
Parameters for the fiber and liquid used at room temperature [39]

	ρ g/cm^3	γ_{lg} mN/m	γ_{fg} mN/m	n_i	ε_i	θ
PP	0.905	–	29.4	1.490	2.2	86°
Water	0.998	72.75	–	1.333	80	86°

B_1 is chosen as 3.72. The constants k_1 and k_2 are calculated by Equation (4) as: $k_1 = 1.34$, $k_2 = 12.0$ (sample 1); $k_2 = 14.2$ (sample 2); $k_2 = 20.1$ (sample 3).

Results of the experiments and simulations are in considerable accord. With the yarn count kept constant, total surface area of a unit length of the yarn increases with the decrease of filament diameter in the yarn. This, in turn, causes an increase first in the interaction area of fiber/liquid and then in the interaction energy, resulting in a higher wettability of the yarn.

Both experiments and simulations show that traveling height rises substantially at first and then slows down, asymptotically approaching a plateau where the effect of the gravity of the liquid column cannot be ignored. This justified the introduction of the gravity term into the expression for Hamiltonian.

Then we followed with a simulation to clarify the influence of fiber orientation on the dynamics of liquid wetting/wicking in a fibrous mass, using the approach we have so far been describing. Fiber declination β_d from the vertical axis was varied in steps of 10°, so the simulation was carried out for 11 different fibrous systems with $\beta_d = 0°$, 10°, ...90°, plus $\beta_d = 45°$. This parametric study for different β_d angles

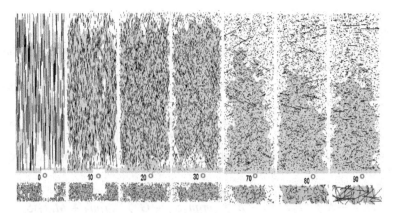

Figure 3. Wetting patterns (vertical and horizontal cross sections) of a fiber mass at different β_d (angles of fiber declination) after 600 Monte Carlo steps.

was intended to show the extent to which this approach could be used to describe fiber systems with different structural features (orientations, for example).

Results are provided in Fig. 3, where each vertical pair of pictures represents the side and top views of the wetting patterns for a cross section of the fibrous mass at a given β_d. Horizontal cross sections are all cut at the distance of 100 cells from the liquid surface.

Two extreme behaviors can be seen. One is vertical to the $W-L$ plane ($\beta_d = 0°$), where the ascending liquid moves at the highest rate but is mostly dispersed. The other is parallel to the $W-L$ plane ($\beta_d = 90°$), and appears to be the lowest in wetting rate, but reveals a liquid pattern that is most heavily aggregated.

3. INTERFACE BETWEEN FIBER AND RESIN IN COMPOSITE MATERIALS

3.1. Model description

Adaptation of the model from liquid transport process to solid adhesion phenomena requires some change in defining the Hamiltonian expression. Let the single-fiber pull-out process be an example. Target system studied in this paper is made of a matrix with a fiber embedded inside. In the case of a 2-dimensional Ising model, the plane that comprises the fiber and matrix can be regarded as a lattice made of a number of square cells. For the sake of simplicity, we use a lattice containing only 5×7 cells as shown in Fig. 4(a), and assume that the sides of a cell are equal to the diameter of the fiber.

To simplify the operation, the lattice is divided in such a way that each cell, if not empty, is filled either with fiber or matrix, i.e., fiber and matrix do not coexist in any single cell. As a result, four possible variables are used to describe the state of each cell:

i) F_i is designated as 1 or 0, to represent the cell i that is either occupied by fiber or not.

ii) f_i is designated as 1 or 0, to represent the fiber cell i in which fiber is either still bonded to the matrix or not.

iii) M_i is designated as 1 or 0, to represent the cell i that is either occupied by matrix or not.

iv) m_i is designated as 1 or 0, to represent the matrix cell i in which matrix is either still bonded to the bulk matrix or not (matrix failure).

The overall energy of the system should be the summation of the energy of each single cell:

$$H = -A \sum_{j}^{cn} F_i F_j S_{ij} - B \sum_{j}^{cn} m_i m_j S_{ij} - G \sum_{j}^{ct} (f_i m_j + M_i f_j) S_{ij} \qquad (5)$$

The above equation shows that all three possible interactions in the system are considered:

i) The cohesive energy between the connected fiber cells, as shown in the first term in the right hand side of Equation (5), where A represents the unit cohesive energy within the fiber, cn above the summation symbol denotes a summation of F values over all the fiber cells in direct connect with cell i, and S_{ij} represents the interaction area between cell i and cell j.

ii) The cohesive energy between the connected matrix cells, as shown in the second term, where B corresponds to the unit cohesive energy within the matrix.

iii) The adhesion interactions between the contacted matrix and fiber cells, as shown in the last term, where G reflects the unit adhesion energy between fiber and matrix, and ct above the summation symbol denotes a summation of m or f values over all the cells in contact with cell i.

As the 2-dimensional pull-out model is the projection of a real 3-dimensional single fiber composite, and assuming that the fiber is in the shape of a perfect cylinder, the interaction area between a fiber cell i_1 and a matrix cell i_2, as shown in Fig. 4(a), should be a half cylindrical surface with both its diameter and height equal to the fiber diameter d. And the interaction area between fiber cell i_1 and fiber cell i_3 is a circle with diameter d. Thus

$$S_{ij} = \begin{cases} \frac{\pi d^2}{2} & \text{(cell } i \text{ is above or under cell } j) \\ \frac{\pi d^2}{4} & \text{(cell } i \text{ is connected with cell } j) \end{cases} \qquad (6)$$

The work done to the system by tensile load W_T is the product of tensile force T_f and the elongation of the fiber. For each fiber cell of length d, the elongation is in the increment of 5% or $0.05d$. This relatively small value should make fiber Poisson's contraction negligible in this study.

Next, for a single fiber cell, as friction is effective only after debonding between fiber and matrix occurs, friction and adhesion interactions are mutually exclusive.

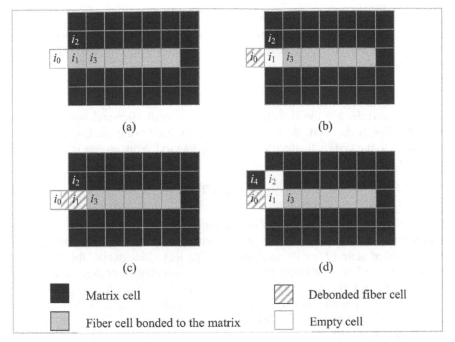

Figure 4. Two-dimensional Ising model in the pull-out process. (a) Initial stage. (b) Fiber breaks during pull-out process. (c) Fiber elongates. (d) Matrix is pulled out by the fiber.

The friction considered here comes only from the shrinkage stress of the matrix to the fiber and is assumed to be constant. As the pull-out process is usually performed at a very slow and even rate, the dynamic effect can be excluded, and friction energy per unit area F_r can be regarded as constant [37]. Then the work W_F done to the system by friction is equal to the product of F_r and friction contact area, and friction contact area is the product of the number of debonded yet still unbroken fiber cells n and the contact area between each fiber cell and surrounding matrix πd^2.

3.2. Monte Carlo simulation

In the simulation, whether a fiber cell will break/debond, or whether a matrix cell will be pulled out from the bulk matrix is determined by the Monte Carlo simulation based on the energy difference before and after the cell state changing (i.e., from a bonded state to a debonded state). To accommodate fiber displacement during the pull-out test, an extra cell i_o is set up as shown in Fig. 4, where its initial state is $F_o = 0$. And F_o will increase until the completion of the test, indicating the fiber displacement.

3.2.1. Energy difference before and after fiber breaking

Fiber breaking at a given cell i is marked by the cell changing from a fiber-occupied state ($F_i = 1$) to an empty state ($F_i = 0$), as shown in Fig. 4(b). The energy difference as a result is:

$$\Delta E = H_2 - H_1 - W_T + W_F \tag{7}$$

where H_2 and H_1 are the Hamiltonians of the system after and before the fiber breaking, W_T is the work done to the system by the tensile load, and W_F is the work done to the system by the friction. According to the discussion in the previous section, Equation (7) becomes:

$$\Delta E = H_2 - H_1 - 0.05T_f \cdot d + \pi d^2 n F_r \tag{8}$$

3.2.2. Energy difference before and after fiber debonding

A debonding fiber cell i is the one that has changed from a bonded state ($f_i = 1$) to a debonded state ($f_i = 0$), as shown in Fig. 4(c). Meanwhile, the fiber cell i_1 is also elongated into the empty cell i_0 ($F_0 = 0$), which then changes into a partial fiber cell ($F_0 = 0.05$) to reflect the elongation of this fiber cell during debonding. Also with the debonding of each successive fiber cell, a value of 0.05 is accrued to the F_0 of each partial fiber cell. This energy difference can also be calculated using Equations (5) and (8).

3.2.3. Energy difference before and after the pull-out of the matrix cell

That the matrix in cell i is pulled out is represented by a cell changing from a matrix-occupied state ($M_i = 1, m_i = 1$) to an empty state ($M_i = 0, m_i = 0$), as shown in Fig. 4(d). The energy difference is:

$$\Delta E = H_2 - H_1 - W_F \tag{9}$$

With the total energy difference for the cell i, the probability W for a change of the system from a state with energy E_1 to a state with energy E_2 is decided in a similar way as described in Section 2.2.

In this study, a realistic macro-system is regarded as a discrete assembly composed of micro-cells with interactions. Accordingly, the failure process of the bulk system is the outcome of a set of failure behaviors of the constituent single cells. For each of these failure behaviors, i.e., fiber breaking, fiber debonding, and matrix yielding, the three ΔEs in Equations (7), (8) and (9) should be examined separately for each cell.

The whole failure process starts with fiber debonding, from the entrance fiber cell to the last fiber cell embedded in the matrix. For each cell, the tensile load T_f needed to make a fiber cell debond or break is determined in a trial algorithm: First set T_f as 0, then T_f is increased steadily with a small increment of dT_f ($= 1$) in each iteration until either fiber debonding or breaking occurs. Then, for the same cell, whether a contacted matrix cell will be pulled out with the fiber is determined.

During the process of fiber debonding, it is the adhesion interaction between the fiber and matrix that pulls the matrix cell out from the bulk matrix. Therefore, when Equation (9) is used to compute the energy difference, W_F should be equal to 0, as adhesion interaction and friction are mutually exclusive.

Upon debonding of the last fiber cell from the matrix, or upon breaking of any of the fiber cells during the debonding process, the post-debonding frictional sliding begins. During this process, only two of the failure modes, i.e., fiber breaking and matrix being pulled out, are possible. Now, as the interface between fiber and matrix has totally failed, i.e., no adhesion interaction exists, W_F then takes effect, and is equal to the product of F_r and the contact area between a single matrix cell and the fiber.

3.3. Results and discussion

This study centers on a computer simulation intended to clarify the pull-out process of a fiber from a bulk matrix. As can be seen from Equations (5) to (9), six parameters are involved in deciding the pull-out behavior: 1) fiber cohesion coefficient A, 2) matrix cohesion coefficient B, 3) interfacial adhesion coefficient G, 4) F_r, friction energy per unit area, 5) the embedded length of fiber in the matrix l, and 6) the fiber diameter d. As these parameters in the energy expression correspond to the characteristics of the materials involved, parametric studies of the pull-out process can be carried out by adjusting values of these coefficients for their effects on the system behaviors.

To perform such a parametric study, one of the parameters is varied while the rest are kept constant. A meaningful parametric study requires that all parameters in the model are independent of, or only weakly dependent on, each other. Usually, the bonding strength (G) can be altered by fiber surface treatments with little influence on the strengths of both fibers (A) and matrix (B) in the system. Also, there is little contact between A and B, and either l or d is obviously an independent parameter. Finally the friction energy F_r is independent of either A or B which represent only the strengths of materials. As to F_r and G, they have been treated as mutually exclusive.

To give a demonstration, only two of the parameters, G and F_r, need to be discussed as examples in the following sections. For generality, quantitative values and their corresponding dimensions can be determined by calibrating against actual experimental tests. In the present study, all physical parameters are in relative and dimensionless terms.

3.3.1. Effect of interfacial bonding strength on the pull-out behavior
Designed to measure the interfacial bonding/debonding properties of fiber reinforced composites, the pull-out tests have been employed extensively to examine the influence of the interfacial adhesion or fiber-matrix bonding strength on the composite behavior.

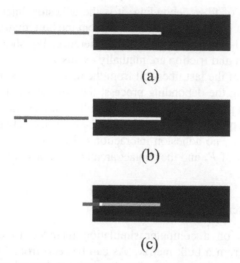

Figure 5. Pull-out process for samples with different interfacial bonding strengths. (a) Fiber is pulled out. (b) Fiber is pulled out with matrix. (c) Fiber breaks before debonding terminates.

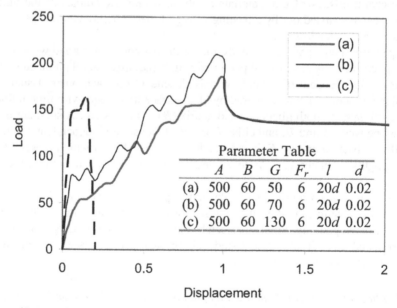

Figure 6. Simulation results of load vs. fiber displacement at various bonding strengths (all parameters are dimensionless).

For three sets of the total parameters with different values of G but constant values for the rest, the simulated pull-out behaviors are shown in Fig. 5 and Fig. 6: (a) $A = 500$, $B = 60$, $G = 50$, $F_r = 6$, $l = 20d$, $d = 0.02$; (b) $A = 500$, $B = 60$, $G = 70$, $F_r = 6$, $l = 20d$, $d = 0.02$; (c) $A = 500$, $B = 60$, $G = 130$, $F_r = 6$, $l = 20d$, $d = 0.02$. In Fig. 5 as well as in the rest of the simulation, the lattice for each case is chosen to contain 80×11 cells.

For the whole process of a fiber debonding and its being pulled out from a matrix, the typical curve of the load applied versus displacement, with parameters cited in case (a), comprises two parts. First there is an increase of tensile load that will propagate the debonded area along the interface and resist the friction in the increased debonding area. And the required tensile load reaches its peak when the whole fiber is debonded from the matrix. The tensile load drops, and then decreases steadily when the fiber is being pulled out against the friction between the fiber and matrix. The trend of the curve agrees with results reported in the literature [36, 37]. In this and later sections, the steadily decreasing part of the curves corresponding to the post-debonding friction sliding process is somewhat truncated, so that the part of the debonding process can be highlighted in greater details.

As a stochastic approach, the Monte Carlo simulation is capable of depicting the stochastic nature of the debonding and pull-out behaviors in a real sample where the interface could hardly be viewed as perfect. As shown in Fig. 6, the debonding curves of the load vs. displacement for the pull-out process, far from being smooth, manifest many jerks, which is as expected and is in accordance with reported results [37]. This reflects the irregular property distribution of the interface between fiber and matrix, justifying the use and demonstrating the power of this stochastic approach (i.e., Monte Carlo simulation) to predicting polymer-fiber interfacial behaviors in the study of composite materials.

Figure 6 indicates that, with the increase of interfacial bonding strength G, the tensile load required to pull the fiber out increases during the debonding period. When G is large enough, as shown in case (c), the fiber breaks before debonding terminates. The rest two curves coincide in the post-debonding area as the friction is identical in both cases.

3.3.2. Effect of friction on the pull-out behavior

For different values of F_r, the pull-out behavior is shown in Fig. 7 and Fig. 8: (a) $A = 500$, $B = 60$, $G = 50$, $F_r = 6$, $l = 20d$, $d = 0.02$; (b) $A = 500$, $B = 60$, $G = 50$, $F_r = 9$, $l = 20d$, $d = 0.02$; (c) $A = 500$, $B = 60$, $G = 50$, $F_r = 11$, $l = 20d$, $d = 0.02$.

The effect is that with the increase of friction, the tensile load required to pull out the fiber and to resist the post-debonding friction increases because, for the whole system, friction works during both the debonding and post-debonding processes, unlike what happens to individual cells. Also, the fiber breaks before it is pulled out, as in case (b), or even breaks before it is entirely debonded from the matrix, as in (c), when friction is high enough.

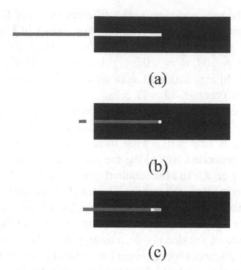

Figure 7. Simulation of pull-out process for samples with different interfacial friction. (a) Fiber is pulled out. (b) Fiber breaks before being pulled out. (c) Fiber breaks before debonding terminates.

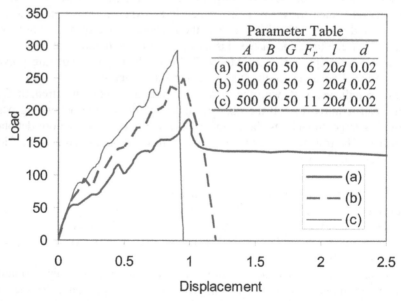

Figure 8. Simulation results of load vs. fiber displacement for different interfacial friction (all parameters are dimensionless).

4. CONCLUSION

The Ising model combined with Monte Carlo simulation is applied in two seemly different interfacial phenomena: liquid transport in fibrous materials and adhesion/debonding between fiber and resin in a composite. These two processes share common ground in that both the wetting and adhesion/debonding processes result from interfacial interactions between two phases (liquid/fiber or fiber/resin) and from cohesion within each phase (cohesion within a liquid or cohesion within a resin block). In both cases, the computational simulations agree with experimental observations, indicating the prospect of applying a simple binary model in similar wetting and adhesion processes.

This 2-D model can be easily expanded, if necessary, into a 3-D one, where the number of the nearest-neighbor cells increases from 8 to 26. A discrete simulation approach, by making changes in lattice size, can be accommodated to material systems on different scales. In other words, this meso-scale approach provides a bridge between our knowledge of the macro-behavior of a material and that of the interactions between its micro-constituents. Since the continuum mechanics approach is incapable of describing behaviors of materials on scales smaller than tens of micrometers, this meso-scale approach is especially useful in the study of materials of a characteristic length scale and in the range of nanos and micros.

REFERENCES

1. R. J. Good, *J. Adhesion Sci. Technol.* **6**, 1269 (1992).
2. P. G. de Gennes, *Rev. Modern Phys.* **57**, 827 (1985).
3. E. W. Washburn, *Phys. Rev.* **17**, 273 (1921).
4. F. W. Minor, *Textile Res. J.* **29**, 931 (1959).
5. F. W. Minor, *Textile Res. J.* **29**, 941 (1959).
6. S. Chwastiak, *J. Colloid Interface Sci.* **42**, 298 (1973).
7. C. J. Nederveen, *Tappi J.* **77**, 174 (1994).
8. R. N. Ibbett and Y. L. Hsieh, *Textile Res. J.* **71**, 164 (2001).
9. D. R. Schuchardt and J. C. Berg, *Wood Fiber Sci.* **23**, 342 (1991).
10. A. Marmur, *Langmuir* **19**, 5956 (2003).
11. F. Brochard, *J. Chem. Phys.* **84**, 4664 (1986).
12. C. Bauer, T. Bieker and S. Dietrich, *Phys. Rev. E* **62**, 5324 (2000).
13. T. Bieker and S. Dietrich, *Physica A* **259**, 466 (1998).
14. G. McHale and M. I. Newton, *Colloid Surfaces A* **206**, 79 (2002).
15. A. V. Neimark, *J. Adhesion Sci. Technol.* **13**, 1137 (1999).
16. D. Quere, *Annu Rev. Fluid Mech.* **31**, 347 (1999).
17. S. Yoshikawa, K. Ogawa, S. Minegishi, T. Eguchi, Y. Nakatani and N. Tani, *J. Chem. Eng Jpn* **25**, 515 (1992).
18. K. Ghali, B. Jones and J. Tracy, *Textile Res. J.* **64**, 106 (1994).
19. N. Mao and S. J. Russell, *J. Appl. Phys.* **94**, 4135 (2003).
20. S. S. Manna, H. J. Herrmann and D. P. Landau, *J. Stat. Phys.* **66**, 1155 (1992).
21. A. Lukkarinen, K. Kaski and D. B. Abraham, *Phys. Rev. E* **51**, 2199 (1995).
22. D. Lukas, E. Glazyrina and N. Pan, *J. Textile Inst.* **88**, 149 (1997).
23. D. Lukas and N. Pan, *Polym. Composites* **24**, 314 (1993).

24. W. Zhong, X. Ding and Z. L. Tang, *Textile Res. J.* **71**, 762 (2001).
25. W. Zhong, X. Ding and Z. L. Tang, *Acta Phys.-Chim. Sinica* **17**, 682 (2001).
26. W. Zhong, X. Ding and Z. L. Tang, *Textile Res. J.* **72**, 751 (2002).
27. X. Ji, Y. Dai, B. L. Zheng, L. Ye and Y. W. Mai, *Composite Interfaces* **10**, 567 (2003).
28. A. K. M. Masud and A. K. M. K. Bin Zaman, *J. Mater. Proc. Technol.* **172**, 258 (2006).
29. J.-K. Kim and Y. W. Mai, *Engineered Interfaces in Fiber Reinforced Composites*, Elsevier, Amsterdam (1998).
30. C. H. Hsueh, *Mater. Sci. Engi. A* **123**, 1 (1990).
31. J. Brandstetter, K. Kromp, H. Peterlik and R. Weiss, *Composites Sci. Technol.* **64**, 65 (2004).
32. W. Sun and F. Lin, *J. Thermoplastic Composite Mater.* **14**, 327 (2001).
33. W. Beckert and B. Lauke, *Composites Sci. Technol.* **57**, 1689 (1997).
34. H. Stang and S. P. Shah, *J. Mater. Sci.* **21**, 953 (1986).
35. J. K. Kim, C. Baillie and Y. W. Mai, *J. Mater. Sci.* **27**, 3143 (1992).
36. C. Marotzke and L. Qiao, *Composites Sci. Technol.* **57**, 887 (1997).
37. C. Delfolie, C. Depecker and J. M. Lefebvre, *J. Mater. Sci.* **34**, 481 (1999).
38. J. N. Israelachvili, *Intermolecular and Surface Forces*, Academic Press, London (1991).
39. D. W. v. Krevelen, *Properties of Polymers: Their Correlation with Chemical Structure, Their Numerical Estimation and Prediction from Additive Group Contributions*, Elsevier, Amsterdam (1990).

Contact Angle, Wettability and Adhesion, Vol. 5, pp. 191–205
Ed. K.L. Mittal
© VSP 2008

Adhesion of hydrophobizing agents: A comparison of values determined by contact angle and the JKR approaches

JUHA LINDFORS,* JANNE LAINE and PER STENIUS

Helsinki University of Technology – TKK, Laboratory of Forest Products Chemistry, P.O. Box 6300, FI-02015 TKK, Finland

Abstract—The work of adhesion of hydrophobizing agents of the type typically used for paper and board – alkenyl succinic anhydride (ASA) and alkyl ketene dimer (AKD) – as well as of poly(dimethylsiloxane) (PDMS) to hydrophilic and hydrophobic model surfaces was studied. The work of adhesion was determined by contact angle measurements and also by a recently developed contact mechanics technique based on the JKR methodology. The results obtained by these different techniques were compared with each other and the correlation between the contact angle and the JKR methods was found to be good. Both the contact angle and the JKR approaches are applicable to determining adhesion to macroscopic surfaces and they also facilitate the investigation of many different materials, which makes them very practical in studying interactions that are of great interest in different applications related to papermaking.

Keywords: Adhesion; contact angle; contact mechanics technique; hydrophobizing agents; alkenyl succinic anhydride; alkyl ketene dimer; contamination of process equipment; papermaking.

1. INTRODUCTION

In papermaking, the contamination of process equipment due to the formation of hydrophobic deposits is a common problem associated with the use of reactive hydrophobizing agents (sizes), such as alkenyl succinic anhydride (ASA) and alkyl ketene dimer (AKD) (for the general molecular structures of ASA and AKD see Fig. 1) [1–5]. Among the factors having an influence on this contamination problem and on attempts to minimize them, surface interactions between hydrophobizing agents and process equipment are of great significance. Spreading and adhesion of ASA on both hydrophilic and hydrophobic surfaces has been studied recently [6, 7]. However, the fact that the softening temperature of these sizes varies from approximately 5°C up to above 60°C limits the applicability of contact angle

---

*To whom correspondence should be addressed. Tel.: +358 9 451 5657; Fax: +358 9 451 4259; e-mail: juha.lindfors@tkk.fi

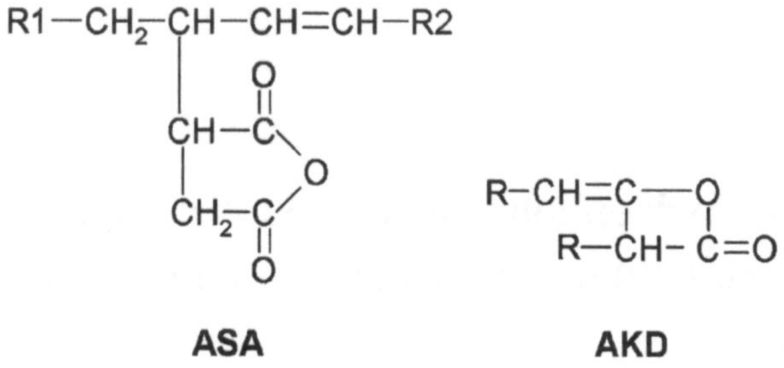

Figure 1. General molecular structures of ASA and AKD. R, R1 and R2 are alkyl chains.

measurements. Also, the tendency of reactive sizes to hydrolyze in aqueous systems, especially the rapid hydrolysis of ASA [8], can present problems when determining the work of adhesion through contact angle measurements (e.g. due to the impossibility to measure interfacial tensions reliably). Hence, there is a need for other methods, in addition to contact angle measurements, to determine the work of adhesion between hydrophobizing agents and different materials.

Recent development of experimental methods [9–11] based on the JKR theory of contact mechanics [12] provides the possibility to measure the work of adhesion between solids. To meet the requirements of the JKR theory (i.e. having an elastic contact and being able to determine the contact area) cross-linked poly(dimethylsiloxane) (PDMS) caps have been commonly used for measurements. However, modified caps have also been developed [10, 13], which allows studies of adhesion properties between materials more relevant to practical applications. Contact mechanics and contact angle methods to determine the work of adhesion have been earlier studied and compared [14–16] and rather good correlation between these two approaches has been often found. This is important when methods complementary to contact angle studies are considered.

This paper presents the application of a method based on the JKR theory, the Micro Adhesion Measurement Apparatus (MAMA) [17], to study the adhesion between hydrophobizing agents and model surfaces. The work of adhesion was also determined by contact angle measurements and the results obtained through MAMA and contact angle measurements were compared. The adhesion properties of PDMS used as a reference material were also measured. The introduction of this contact mechanics technique for measuring the adhesion of hydrophobizing agents is expected to facilitate a more thorough understanding of paper machine contamination and a better internal consistency of adhesion measurements.

Also, it should be mentioned that such well-established methods as Atomic Force Microscopy (AFM) and surface force apparatus (SFA) have been widely used for measuring adhesion between solid materials [18, 19]. However, when desiring to examine adhesion on a macroscopic scale to materials present in paper machines, these methods are not directly applicable and thus they were not considered in this study.

2. EXPERIMENTAL

2.1. Materials

Poly(dimethylsiloxane) (PDMS) used was a two-component system supplied by Dow Corning Corporation (USA). According to the manufacturer the PDMS was solvent-free and its density was 1.03 g/cm^3. Alkenyl succinic anhydride (ASA) in liquid form for contact angle measurements was supplied by Ciba Specialty Chemicals, Raisio, Finland. The same ASA was used earlier and is described in detail elsewhere [6]. The purity of the ASA was ca. 99%. Solid ASA, which was used to prepare surfaces for MAMA measurements, was supplied by Kemira Chemie GmbH, Krems, Austria. A commercial alkyl ketene dimer (AKD) product from Ciba Specialty Chemicals, Raisio, Finland was used for contact angle measurements. The purity of the AKD was >85%, impurities being mainly residual olefin chains from the synthesis by dimerization of aliphatic acid halides, based mainly on stearic acid (C18), through dehydrohalogenation. Its melting point was ca. 15°C and its density was 0.875 g/cm^3. AKD with a higher melting point (ca. 62°C), also from Ciba Specialty Chemicals, was used for preparing model surfaces for MAMA measurements. The AKD was purified by recrystallizing it twice from acetone. Milli-Q water was used throughout the investigation.

Hydrophilic surface: Smooth silicon wafers were cleaned by immersing them into alkaline and acidic hydrogen peroxide solutions (first 10 min at 75–85°C in alkaline, then 10 min at 75–85°C in acidic solution) and then rinsed successively with water and ethanol.

Hydrophobic surface: This was prepared by methylating the cleaned and dried SiO$_2$ surface by immersing it in the 0.05% dichlorodimethylsilane in xylene for 2 h (for more details see Lindfors *et al.* [6]).

2.2. Preparation of PDMS caps

PDMS base and cross-linker were thoroughly mixed in a 10 : 1 (parts per weight) ratio after which the mixture was deaerated in vacuum for 1 h. Then hemispherical droplets of solution were formed on a smooth substrate and cured at 105°C for 60 min. Hexane extraction was performed in the manner described by Rundlöf *et al.* [11]. Caps with a ca. 1 mm radius of curvature were obtained.

2.3. Preparation of ASA- and AKD-coated caps

ASA and AKD were spin coated on the PDMS caps from a 0.5% (w/v) hexane solution at a spin rate of 2700 rpm.

2.4. Measurement of contact angles and surface tensions and calculation of the work of adhesion

A CAM 200 contact angle goniometer (KSV Instruments Ltd, Helsinki, Finland) was used for determination of contact angles and liquid surface tensions. For experimental details and methods for determination of surface tensions (pendant drop profile fitting) and contact angles see Lindfors *et al.* [6]. From liquid surface tensions and contact angle measurements, the work of adhesion, W_a, was calculated using the Young-Dupré equation as:

$$W_a = \gamma_{LV}(1 + \cos \theta) \qquad (1)$$

where γ_{LV} is the liquid surface tension and θ is the contact angle.

2.5. Micro adhesion measurement apparatus

The work of adhesion was also measured using a Micro Adhesion Measurement Apparatus, MAMA (Akribi Kemikonsulter, Sundsvall, Sweden), which is based on the JKR theory. Details about instrumentation and its theoretical background are given by Rundlöf and Wågberg [17]. An experimental protocol similar to that employed by Rundlöf and Wågberg was used for establishing approach, contact and separation of the two bodies. As emphasized by Ghatak *et al.* [20], the values obtained through contact mechanics measurements, especially adhesion hysteresis, are time-dependent to varying degrees, depending on interfacial chemistry. Here a one minute delay was used between each loading and unloading step. This interval, which was considerably shorter than that used e.g. by Rundlöf and Wågberg [17], was selected to avoid very large hysteresis. One minute was also found to be enough for the load and contact area to be quasi-stabilized. A maximum load of 250–300 mg was used. All MAMA measurements were performed in laboratory air ($T = 24 - 26°C$, RH ca. 45%).

2.6. Atomic force microscopy (AFM)

AFM imaging was performed using a NanoScope IIIa Multimode scanning probe microscope (Digital Instruments, Inc. Santa Barbara, CA, USA). The surfaces were scanned in the tapping mode in air using commercial Si cantilevers (Digital Instruments) with a resonance frequency of about 300 kHz.

2.7. Infrared spectroscopy

ATR-FTIR measurements were performed with a Bio-Rad FTS 6000 spectrometer using a diamond ATR microcrystal. The method is described in more detail by Vikman and Vuorinen [21].

3. RESULTS

3.1. Surface composition of ASA and AKD coated PDMS caps

PDMS caps coated with ASA and AKD were analysed with ATR-FTIR and AFM in order to verify successful preparation of surface films from these hydrophobizing agents. ATR-FTIR spectra of uncoated PDMS, ASA, and PDMS coated with ASA are presented in Fig. 2. Corresponding results for AKD are shown in Fig. 3.

The ATR-FTIR analysis depth is ca. 1 μm. The PDMS underneath the surface layer can be recognized in the spectra of ASA and AKD on PDMS. However, peaks typical of ASA and AKD at ca. 2915, 2850 and 1780 cm^{-1} can also be clearly identified on the coated PDMS caps. From ellipsometry measurements performed

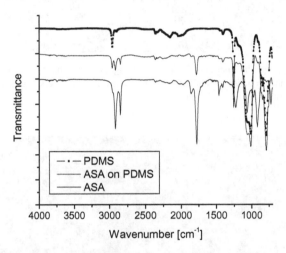

Figure 2. ATR-FTIR spectra of PDMS, PDMS coated with ASA and ASA (top to bottom). For clarity, the ASA on PDMS and ASA spectra have been shifted downwards.

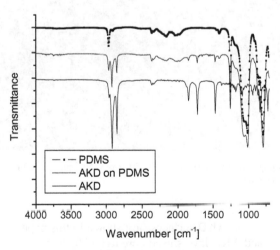

Figure 3. ATR-FTIR spectra of PDMS, PDMS coated with AKD and AKD (top to bottom). For clarity, the AKD on PDMS and AKD spectra have been shifted downwards.

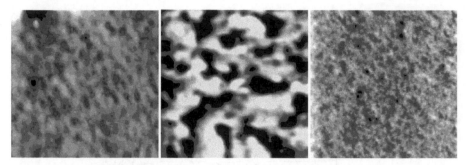

Figure 4. AFM topography images of PDMS (left image, rms-roughness = 1.30 nm), ASA (middle image, rms-roughness = 6.69 nm), and AKD (right image, rms-roughness = 0.90 nm) surfaces (in all images: size = 5 μm × 5 μm, z = 14 nm).

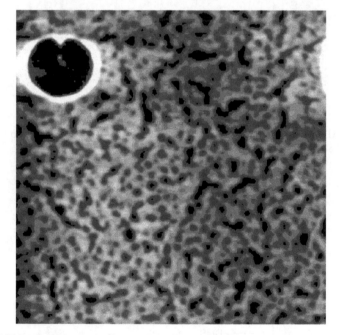

Figure 5. AFM topography image (image size: 50 μm × 50 μm, z = 40 nm) of ASA on PDMS (rms-roughness = 8.93 nm).

on spin coated ASA and AKD surfaces (not shown in this paper), we evaluated the thickness of ASA and AKD layers used here to be 20–30 nm.

AFM topography images (of size 25 μm^2) recorded from PDMS as well as from ASA and AKD coated PDMS are shown in Fig. 4. When PDMS was coated with ASA, a more pronounced topography change was introduced than when using AKD.

In fact, in the case of AKD coated PDMS the rms-roughness of the surface even slightly decreased (from 1.30 nm to 0.90 nm). The rms-roughness of ASA on PDMS was 6.69 nm and when determined from 50 μm × 50 μm image (shown in Fig. 5) the rms-roughness was 8.93 nm. The top left corner of Fig. 5 shows a defect in the ASA layer. Similar defects, with diameters about 10 μm, were occasionally present on ASA surfaces. However, we conclude that the combination of different analyses of the surfaces indicates that good surface coverage was attained with both ASA and AKD and that the surface layers are sufficiently smooth for contact mechanics studies.

3.2. Surface tensions

The liquid surface tensions, determined using pendant drops, of PDMS, ASA, and AKD in ambient air are given in Table 1. Surface tensions were recorded at one second intervals for a 30 s period during which the values were fairly stable. The values given in Table 1 are averages (during the 30 s) and standard deviations are based on five measurements.

3.3. Contact angles

The kinetics of spreading of ASA on both hydrophilic and hydrophobic surfaces has been described previously (Lindfors *et al.* [6]). Spreading of PDMS, and AKD was rather similar to ASA. The same number of measurements (five) were performed on each surface and contact angles used for calculations of the work of adhesion were taken at $t = 1$ min. The contact angles are presented in Table 2.

Table 1.
Surface tensions of PDMS, ASA and AKD (at 24°C)

Material	Surface tension, mN/m
PDMS	21.5 ± 0.8
ASA	33.2 ± 0.5
AKD	29.0 ± 1.0

Table 2.
Contact angles of PDMS, ASA and AKD ($t = 1$ min)

| Material | Contact angle, degrees | |
	On the hydrophilic surface	On the hydrophobic surface
PDMS	25 ± 2	25 ± 2
ASA	25 ± 2	47 ± 2
AKD	21 ± 3	53 ± 3

198   J. *Lindfors* et al.

3.4. Contact mechanics measurement of adhesion

The work of adhesion, W_a, between two PDMS caps was measured in order to be able to compare the results with those previously obtained [11, 22] for the same material system. By fitting the measured values (see Fig. 6) of the cube of the contact area radius as a function of the load to the JKR theory according to the following equation

$$a^3 = \frac{R}{K}\left(P + 3\pi W_a R + \sqrt{6W_a R P + (3\pi W_a R)^2}\right) \qquad (2)$$

(where a = radius of the contact area between the two surfaces, R = equivalent radius of curvature of the system, P = applied load, K = elastic constant of the system) the work of adhesion between two PDMS caps, W_a, calculated from the loading data was found to be 44 ± 3 mJ/m^2. This corresponds to a PDMS surface energy of 22 mJ/m^2, which is in good agreement with the surface energy value (21.5 mJ/m^2) determined here with the contact angle measurements and values, 22–24 mJ/m^2, reported in the literature [11, 22].

Figures 7–12 show representative curves obtained from similar experiments for all of the studied material systems, i.e. PDMS-hydrophilic (Fig. 7), PDMS-hydrophobic (Fig. 8), ASA-hydrophilic (Fig. 9), ASA-hydrophobic (Fig. 10), AKD-hydrophilic (Fig. 11) and AKD-hydrophobic (Fig. 12). The values presented here for the work of adhesion were all obtained by fitting the loading data to Eq. (2).

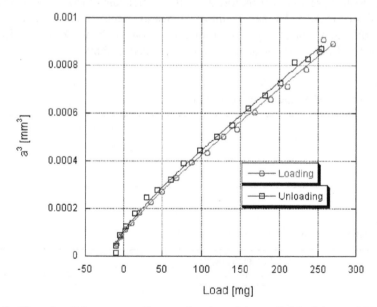

Figure 6. The cube of the contact radius as a function of the applied load for two PDMS caps ($W_a = 44.7 \pm 0.4$ mJ/m^2).

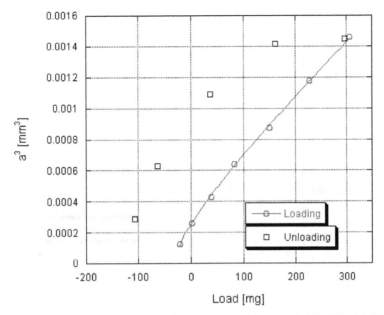

Figure 7. The cube of the contact radius as a function of the applied load for PDMS-hydrophilic SiO$_2$ system ($W_a = 49.2 \pm 0.3$ mJ/m^2).

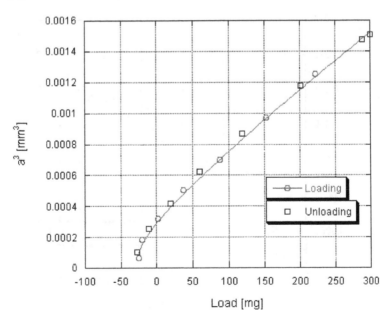

Figure 8. The cube of the contact radius as a function of the applied load for PDMS-hydrophobic SiO$_2$ system ($W_a = 45.1 \pm 0.2$ mJ/m^2).

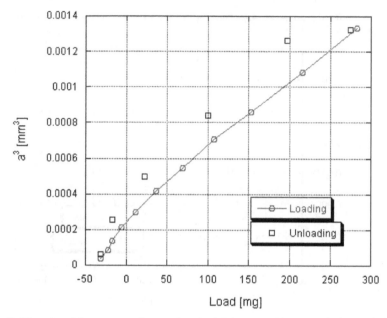

Figure 9. The cube of the contact radius as a function of the applied load for ASA-hydrophilic SiO_2 system ($W_a = 60.9 \pm 0.6$ mJ/m^2).

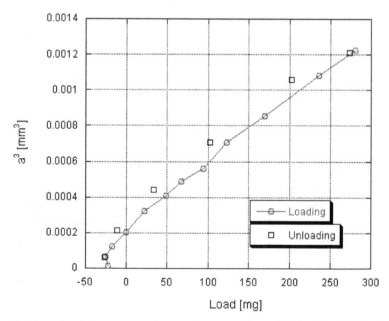

Figure 10. The cube of the contact radius as a function of the applied load for ASA-hydrophobic SiO_2 system ($W_a = 47.9 \pm 1.0$ mJ/m^2).

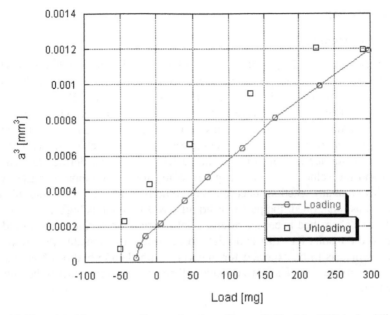

Figure 11. The cube of the contact radius as a function of the applied load for AKD-hydrophilic SiO_2 system ($W_a = 65.1 \pm 0.4$ mJ/m^2).

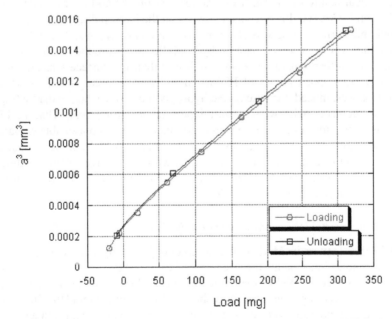

Figure 12. The cube of the contact radius as a function of the applied load for AKD-hydrophobic SiO_2 system ($W_a = 45.0 \pm 0.3$ mJ/m^2).

4. DISCUSSION

4.1. Surface tensions

The surface tension, 21.5 ± 0.8 mN/m, of liquid PDMS, which was used as a reference surface, was in good agreement with the literature data (22–24 mN/m) [22]. The possible effect of the crosslinker used for curing of PDMS caps was also investigated; there was no change at all in the surface tension, whether the curing agent was added or not. This implies that measurements done with the contact mechanics technique should be fully comparable with contact angle measurements. Because the surface tensions of ASA and AKD were stable during the measuring period and the values were in a range expected for hydrophobic surfaces, we do not believe that any reactions occurred that could have affected the values obtained. However, as mentioned, the purity of liquid AKD was not as high as that of ASA. However, if there had been a substantial influence of impurities, one would have expected the results for ASA and AKD to be different. Actually, they were quite similar, which is to be expected for materials with double hydrocarbon chains of about the same length. Thus, we consider all surface tension values to be internally consistent.

4.2. Contact angles

The chemical nature of the surfaces implies that one should expect better wetting by AKD and ASA (i.e. lower contact angles) on the hydrophilic surface than on the hydrophobic surface. This was also observed in measurements. For PDMS, equal contact angles were recorded both on hydrophilic and hydrophobic surfaces. Also, because the surface tension of PDMS is lower than the surface tensions of ASA and AKD, the contact angles of PDMS should be lower than those of ASA and AKD. This was, indeed, the case on the hydrophobic surface, but on the hydrophilic surface all contact angles were, within experimental error, equal. That the PDMS should wet the pure silicon surface well is not unexpected. A reasonable explanation for the low contact angles of ASA and AKD is that their hydrophilic moieties (see Fig. 1) interact through acid/base interactions with the relatively acidic hydrophilic surface. The hysteresis in MAMA adhesion measurements on the hydrophilic surfaces is indicative of such interactions.

4.3. Contact mechanics measurements of adhesion

A distinctive feature in MAMA measurements is the hysteresis observed when measuring adhesion (for all the used materials: PDMS, ASA and AKD) to the hydrophilic surface. Within experimental error, there was no hysteresis in the adhesion to the hydrophobic surface. It should also be kept in mind that in this work, in order to minimize hysteresis, the delay between loading and unloading steps was much shorter than in previous work [11, 23]. Interpenetration, roughness (on molecular and on micro-scale) and formation of chemical linkages between surfaces in contact are examples of reasons previously stated to contribute to hysteresis [11, 22]. At least interpenetration and formation of linkages should be time-dependent. The study of the hysteresis phenomenon was not the main objective of this study and, therefore, is not discussed further here. However, the results (for example, the low contact angles of ASA and AKD) give the impression that chemical reasons would play a key role here. A general summary of previous discussion about hysteresis phenomenon in contact mechanics measurements can be found in a paper by Packham [14].

Table 3.
Work of adhesion of PDMS, ASA and AKD on hydrophilic and hydrophobic surfaces in air

Material	Work of adhesion, mJ/m^2	
	On the hydrophilic surface	On the hydrophobic surface
PDMS	41 ± 2	41 ± 2
ASA	63 ± 2	56 ± 2
AKD	56 ± 2	46 ± 2

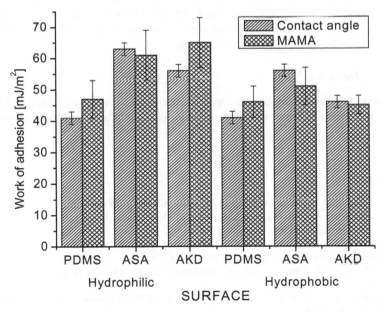

Figure 13. Work of adhesion of PDMS, ASA, and AKD on different surfaces determined from contact angle and MAMA measurements.

J. Lindfors et al.

4.4. Adhesion

The values of the work of adhesion calculated from contact angle measurements (Eq. (1)) and the data in Tables 1 and 2 are presented in Table 3.

The values of the work of adhesion determined from both contact angle measurements and contact mechanics measurements (MAMA) are presented in Fig. 13.

From the results it can be seen that contact angle and MAMA measurements correlate satisfactorily; the results obtained by these two methods agree with each other within experimental error for all six studied systems.

5. CONCLUSIONS

The correlation between contact angle and JKR methods was found to be satisfactory, and thus, the choice of preferred method can be made depending on the system to be investigated. In this worlc, comparison of adhesion values obtained from contact angle and JKR data was the main objective. An interesting objective for further study would be a comparison of the hysteresis observed in both contact angle and JKR measurements.

Both contact angle and JKR techniques are applicable to measuring adhesion to macroscopic scale surfaces and seem to be effective tools for adhesion studies of hydrophobizing agents. Also, because these methods can be easily extended to the studies of other materials similar to ASA/AKD, both the contact angle and the JKR methods will be very useful in studying interactions that are of great interest in different applications related to papermaking.

Acknowledgements

The work was funded by the National Technology Agency of Finland (TEKES). It was performed as a part of "Shine Pro" project in "Clean Surfaces" technology program.

REFERENCES

1. L. Petander, T. Ahlskog and A. J. Juppo, *Paperi ja Puu* **80**(2), 100-103 (1998).
2. J. P. Koskela, O. E. O. Hormi, J. C. Roberts and G. Peng, *Appita J.* **56**(3), 213-217 (2003).
3. D. T. Nguyen, *Tappi J.* **81**(6), 143-151 (1998).
4. S. Knubb and C. Zetter, *Nordic Pulp Paper Res. J.* **17**, 164-167 (2002).
5. L. Neimo, *Papermaking Sci. Technol.* **4**, 151-203 (1999).
6. J. Lindfors, S. Ylisuvanto, T. Kallio, J. Laine and P. Stenius, *Colloids Surfaces, A* **256**, 217-224 (2005).
7. J. Lindfors, S. Ahola, T. Kallio, J. Laine, P. Stenius and M. Danielsson, *Nordic Pulp Paper Res. J.* **20**, 453-458 (2005).
8. J. M. Gess and D. S. Rende, *Tappi J.* **4**(9), 25-30 (2005).
9. M. K. Chaudhury and G. M. Whitesides, *Langmuir* **7**, 1013-1025 (1991).
10. V. S. Mangipudi, E. Huang, M. Tirrell and A. V. Pocius, *Macromol. Symp.* **102**, 131-143 (1996).

11. M. Rundlöf, M. Karlsson, L. Wagberg, E. Poptoshev, M. Rutland and P. Claesson, *J. Colloid Interface Sci.* **230**, 441-447 (2000).
12. K. L. Johnson, K. Kendall and A. D. Roberts, *Proc. Royal Soc. London, Series A* **324**, 301-313 (1971).
13. J. Forsstroem, M. Eriksson and L. Wagberg, *J. Adhesion Sci. Technol.* **19**, 783-798 (2005).
14. D. E. Packham, *Int. J. Adhesion Adhesives* **16**, 121-128 (1996).
15. M. K. Chaudhury, *J. Adhesion Sci. Technol.* **7**, 669-675 (1993).
16. V. Mangipudi, M. Tirrell and A. V. Pocius, *Langmuir* **11**, 19-23 (1995).
17. M. Rundlöf and L. Wågberg, *Colloids Surfaces, A* **237**, 33-47 (2004).
18. H. Butt, B. Cappella and M. Kappl, *Surface Sci. Reports* **59**, 1-152 (2005).
19. J. N. Israelachvili, *Intermolecular and Surface Forces*, 2nd ed., p. 450, Academic Press, New York (1992).
20. A. Ghatak, K. Vorvolakos, H. She, D. L. Malotky and M. K. Chaudhury, *J. Phys. Chem. B* **104**, 4018-4030 (2000).
21. K. Vikman and T. Vuorinen, *J. Imaging. Sci. Technol.* **48**, 138-147 (2004).
22. M. K. Chaudhury and G. M. Whitesides, *Langmuir* **7**, 1013-1025 (1991).
23. A. Falsafi, P. Deprez, F. S. Bates and M. Tirrell, *J. Rheology* **41**, 1349-1364 (1997).

Contact Angle, Wettability and Adhesion, Vol. 5, pp. 207–227
Ed. K.L. Mittal
© VSP 2008

Wettability aspects and the improvement of adhesion of UV curable powder coatings on polypropylene substrates

MARIËLLE WOUTERS[1,*] and PERE CASTELL MUIXÍ[2]

[1] *TNO Science and Industry, de Rondom 1, 5612 AP Eindhoven, The Netherlands*
[2] *Consejo Superior de Investigaciones Científicas, Miguel Luesma Castan 4, 50015 Zaragoza, Spain*

Abstract—Different benzophenone type photo-initiators were photo-grafted onto polypropylene. The polymer surfaces were analysed by means of contact angle measurements, UV spectroscopy and FTIR-ATR. The photo-initiator modified surfaces showed a higher wettability and higher surface energies increasing from 26 mJ/m^2 for the pure polypropylene surface up to 36 mJ/m^2 for the photo-initiator modified surfaces. Different acrylates were subsequently grafted efficiently onto the photo-initiator modified polymer surfaces. FTIR-ATR and contact angle measurements confirmed the presence of the grafted chains. The surface energy of the acrylate grafted samples thus increased up to 70 mJ/m^2 depending on the type of acrylate used. Finally UV curable powder coatings were applied onto the photo-initiator modified polypropylene substrates and the adhesion was found to be improved. Adhesion was measured by the pull-off test and a value of approximately 0.20 MPa was obtained for pure polypropylene while the photo-initiator modified polypropylene reached pull-off value of 2.00 MPa.

Keywords: Surface modification; photo-grafting; contact angle measurements; UV curable powder coatings; adhesion; polypropylene.

1. INTRODUCTION

There has been an upward trend in the amount of plastics being used in industries such as automotive, sporting goods, medical, and electronic. This trend has been influenced by the drive to reduce weight and lower system-costs through design flexibility, increased process efficiency, and increased part performance. Unfortunately, many polymeric materials have a poor tendency to bond to other materials because of their inherent inert chemical structure and thus require a pre-treatment for coating, laminating or bonding.

Chemical or physical modification of the surfaces of polymeric materials is undoubtedly a fascinating field for research as well as a practical way to improve their value.

---

*To whom correspondence should be addressed. Tel.: +31 40 265 0342; Fax: +31 40 265 0302; e-mail: marielle.wouters@tno.nl

Conventional and modern methods of surface modification now provide improved coatability along with the ability to bond plastics which were not capable of being bonded previously [1, 2]. This is especially relevant to low energy surfaces such as polypropylene (PP) and other olefinic materials. Due to their lack of functionalities these materials require a chemical [3] or physical [4] treatment before their final application. This requires tailoring of surface properties of these materials [5, 6]. In industry the main methods to improve adhesion of polyolefins are based on chemical modification of the interfacial polymer chains with polar groups as hydroxyl, carbonyl and carboxylic acid groups [7]. Several methods have been described to modify the surface of the PP, such as the use of block copolymers [8], flame treatment [9], plasma and corona treatments [10–12], and low-pressure glow discharges [13], among others. Such surface modification techniques are difficult to control, and often cause problems regarding uniformity and reproducibility along with another disadvantage i.e. they should be used within a short time between treatment and application of the coating.

Free radical grafting has also been used for surface modification on an industrial scale. This process is performed during extrusion of the material and involves the formation of free radicals along the backbone of the polyolefin [14, 15]. However, such materials often show degradation of mechanical properties due to side reactions such as chain scission. Furthermore, the grafted material is composed of a mixture of modified and pure material and, consequently, the adhesion improvement depends on the surface arrangements and the diffusion of chains. A review by Rätzsch *et al.* [16] discusses all the radical reactions which can occur on a polypropylene surface and illustrates the complexity.

Another possibility is the grafting of functional monomers using UV radiation. This has been widely described in the literature [17–27]. The use of UV radiation offers an excellent alternative to the free radical approach because of its simplicity. Photo-induced grafting is known to be a useful technique for the modification or functionalisation of polymeric surfaces since photo-chemically produced triplet states on carbonyl groups can abstract hydrogen atoms from almost all polymers, thus generating radicals. Therefore, high concentrations of active species can be produced locally at the surface without interfering with the bulk properties of the polymer. In addition to the simplicity of the procedure, the equipment cost is lower for UV irradiation than for ionising irradiation.

This UV technology has been applied to modify the surfaces of various polymers [28, 29] especially polyolefins [17–27] in order to improve their adhesion and wettability.

During the grafting reaction of different monomers onto a polymer surface, the homopolymerisation of the monomer used competes with the grafting, reducing the overall graft reaction yield. The approach proposed by Bowman and coworkers [30] has solved this problem. The graft polymerisation in their study is performed in two steps. In a first step benzophenone abstracts a proton from the surface of the substrate to form semipinacol radicals as well as radicals on the surface itself. These radicals recombine with the surface radicals to form a so-called surface initiator, which is stable and thus the treated material can be stored for future usage. The surface initiators can react in a second step with a monomer solution initiating the polymerisation in the presence of UV light. The formation of homopolymer is reduced since the semipinacol radicals generated have a very short lifetime and prefer to recombine or terminate the growing chains. In a previous paper [31] we had described the photo-grafting of photo-initiators on the surface of PP to increase its surface energy and wettability. The grafted photo-initiators can generate radicals on the surface of the PP upon exposure to UV light. These radicals on the surface of the PP not only increase the wettability of the PP but also generate covalent bonds between the surface and a reactive monomer applied onto the polymer surface.

In this paper we describe the application of UV curable powder coatings onto PP using a two-step process. In the first step the surface of the PP is modified by photo-grafting of different photo-initiators. While in the second step a powder coating is applied and cross-linked using UV light at moderate temperatures. The photo-initiator modified polymer surfaces were analysed by means of contact angle measurements, UV spectroscopy and FTIR-ATR before applying the UV curable powder coating. The final properties of the cured powder coating were also studied, focussing on the adhesion of the powder coating to the photo-initiator modified PP surface.

2. EXPERIMENTAL

2.1. Materials

The polypropylene (PP) sheets (10 cm × 10 cm × 2 mm) used in this study were supplied by Vink Kunststoffen (Didam, The Netherlands). All chemicals used in this study were supplied by Sigma Aldrich Chemie B.V. (Zwijndrecht, The Netherlands) unless otherwise noted. The surface of the PP was cleaned successively with acetone and ethanol before use. For the surface modification reaction, the photo-initiators benzophenone, 4-hydroxybenzophenone, 2,4-dihydroxybenzophenone, 2,3,4-trihydroxybenzophenone, and 2-hydroxy-4-octyloxy-benzophenone were used as received. Dichloromethane was distilled under reduced pressure, while 18-crown-6, potassium carbonate, and ω-bromooctane, were used without further purification in the synthesis of the initiators with aliphatic chains. The monomeric acrylates, acrylic acid, pentaerythritol triacrylate and 2-hydroxyethyl acrylate, were used as received. The binders of the powder coating formulations were based on unsaturated dimethacrylate polyesters that are available from UCB Chemicals under the trademark Uvecoat®. Uvecoat 1000 and Uvecoat 9010 were kindly supplied by UCB Chemicals (Drogenbos, Belgium) and used as received. The photo-initiator in these formulations, Irgacure 651 (IC651), was supplied by Ciba Speciality Chemicals (Groot-Bijgaarden, Belgium) and used without further purification.

2.2. Synthesis

2.2.1. Synthesis of photo-initiators with aliphatic chains

As precursor of the initiator with aliphatic chains the corresponding hydroxyl benzophenone derivative was used. This precursor (5 mmol) was mixed in acetone (50 mL) together with ω-bromooctane (10 mmol, equimolar with respect to the hydroxyl functionality), potassium carbonate (25 mmol), and 18-crown-6 (0.5 mmol). This mixture was heated and refluxed for 24 hr. After this time the reaction was complete as observed by thin layer chromatography using toluene/acetone (9/1) as eluent. The solvent was evaporated and the remaining solid was dissolved in chloroform and washed three times with 100 mL water. The organic phase was evaporated under vacuum to obtain a solid that was re-crystallized from heptane and dried under vacuum. The purity of the modified benzophenone was confirmed by ^{1}H- and ^{13}C-NMR analyses.

2.3. Grafting

2.3.1. Grafting of the photo-initiator

The photo-initiator was dissolved in an appropriate solvent (e.g. methanol, dichloromethane, acetone, or toluene depending on functional groups). The solution was applied onto a PP sheet using a wire bar resulting in a film thickness of 30 μm. The solvent was evaporated and the samples were subsequently irradiated with UV light at 20 cm distance from the source with a high-pressure mercury lamp (H type, 400 W) in a UVA Cube (Dr Hönle AG, Germany), under a nitrogen atmosphere. The irradiation time was varied between 10 s and 600 s and the UV dose was measured using a UV Power Puck radiometer (EIT Inc, USA). The irradiated samples were washed successively with ethanol and acetone and dried overnight. During storage the samples were covered with an aluminium foil to prevent photo-degradation.

2.3.2. Grafting of acrylates

Acrylic acid (AA), pentaerythritol triacrylate (PETA) and 2-hydroxyethyl acrylate (HEA) were grafted onto pure and photo-initiator modified PP sheets at room temperature. 100 μm thick films of the pure acrylates were applied using a wire bar. A quartz plate (2 mm thick) was used to cover the whole sample to ensure a good contact between the acrylate and the polypropylene and also to prevent oxygen inhibition. The grafting was initiated by UV light and exposure time was varied between 30 and 300 s. After irradiation the samples were cleaned (removing excess monomer) using an appropriate solvent, i.e. boiling water in the case of acrylic acid while acetone or ethanol was used for the other acrylates. The samples were dried overnight under vacuum prior to any measurement.

2.4. Powder coating procedure

2.4.1. Powder preparation

The unsaturated polyesters used in this study were from UCB Chemicals available under the trademark Uvecoat®. In our studies we used a mixture composed of two unsaturated polyesters, an amorphous and a semi-crystalline resin. The amorphous resin (Uvecoat 1000) is characterised by a glass transition temperature (T_g) of approx. 47°C and a relatively high viscosity (1000 mPa.s at 200°C). The semi-crystalline resin (Uvecoat 9010) is characterised by a melting point (T_m) of approx. 85°C and a very low viscosity (< 10 mPa.s at 200°C). The two resins were mixed in a proportion described in the literature as the one that offers the best performance, i.e. 75 wt% Uvecoat 1000 and 25 wt% Uvecoat 9010.

In order to mix the two resins with the photo-initiator and some other additives, all the components were ground using a pin mill (Alpine 100 UPZ, Germany) and extruded using a twin-screw extruder (APV Baker, UK) at 80°C and 250 rpm. After extrusion the mixture was ground again and sieved to obtain a powder with a particle size below 150 μm.

2.4.2. Fluidization

The powder coating was placed inside a round recipient (half filled with powder) and compressed dry air at a pressure of 1 Pa was introduced through a porous plate. Under these conditions the powder was fluidised and behaved as a liquid. Polypropylene sheets pre-heated at a temperature of approximately 100°C were dipped in the fluidised powder for a few seconds. Under these conditions coatings with a thickness of 100 μm were obtained after curing.

2.4.3. UV curing

The samples were irradiated in a closed cabinet, UVACube (Dr. Hönle AG, Germany), which was equipped with an H type UV lamp (400 W). The cabinet was equipped with a home-built heating plate that allowed curing at controlled elevated temperatures. The whole compartment can be flushed with air or dry nitrogen to evaluate the influence of the atmosphere on the curing reaction. The UV doses were measured using a UV Power Puck radiometer. All powder coatings investigated in this study were cured at 100°C under a nitrogen atmosphere and were irradiated for 60 s (which corresponds to a UV dose of 1200 mJ/cm^2).

2.5. Characterisation

2.5.1. UV spectroscopy

UV absorption spectra of the photo-initiators were obtained using a Hitachi UV/Vis U-2001 spectrophotometer. The spectra of the pure polypropylene and the photo-initiator modified polypropylene were recorded directly on films of approximately 100 μm thickness.

2.5.2. Contact angle measurements

Contact angles of water and diiodomethane were determined using a goniometer (G10, Krüss) interfaced to image capture software (Drop Shape Analysis 1.0 software). The surface energy was calculated using the values of the contact angles of the water and diiodomethane by the Owens–Wendt method [32].

2.5.3. Contact angle measurements using polymer melts

Surface tension development of a polymer melt, as a function of temperature, was investigated using a conventional goniometer (G10, Krüss). A drop of the polymer melt was formed at elevated temperature on a sample holder in the measurement chamber equipped with a heating element (Linkam, TMS93). During the measurement, pictures of the drop were taken at a fixed time interval using a CCD camera. A drop shape analysis program then fitted the profile of the drop contour and provided the values of the contact angle, the drop volume, and the surface tension. The contact angle was taken on both sides of the drop and was averaged.

Typically, about 10 mg of the polymer was used for the size of the sample holder used in our equipment. After placing the sample in the measurement chamber the temperature was raised at a heating rate of $10°C/min$. A sequence of pictures were acquired (one picture every 30 s) after the polymer was melted and equilibrated. The measurement always started with an isothermal period for 30 minutes after which the temperature was lowered at a cooling rate of $5°C/min$ to the next temperature of measurement. Each temperature of measurement was kept for 30 minutes during which the pictures were taken.

2.5.4. FTIR-ATR

The infrared spectra were recorded with attenuated total reflectance (ATR) accessory using a BioRad FTS 3000MX Excalibur spectrometer with a MTC detector. The spectrometer was equipped with a Golden GateTM Diamond ATR Top-Plate (Specac). Samples of size 20×20 mm were placed between the diamond crystal (refractive index 2.5) and a stainless steel cover with a variable pressure, to ensure good contact between the sample and the crystal. Usually 20 scans with a resolution of $4 \, cm^{-1}$ were collected.

2.5.5. Pull-off measurements

A Universal tensile tester model 112.10kN.L (TesT GmbH, Germany) was used to monitor the force required to detach the coating from the substrate. The tensile tester was controlled with a model 810 control and display electronics (TesT GmbH, Germany).

An aluminium stud (area $0.5 \, cm^2$) was glued to the coating using an epoxy based glue. The glue was dried for 24 hr at room temperature. The coating was sanded slightly to increase its adhesion to the stud. After curing, the excess glue was removed and the coating was cut around the stud until reaching the surface

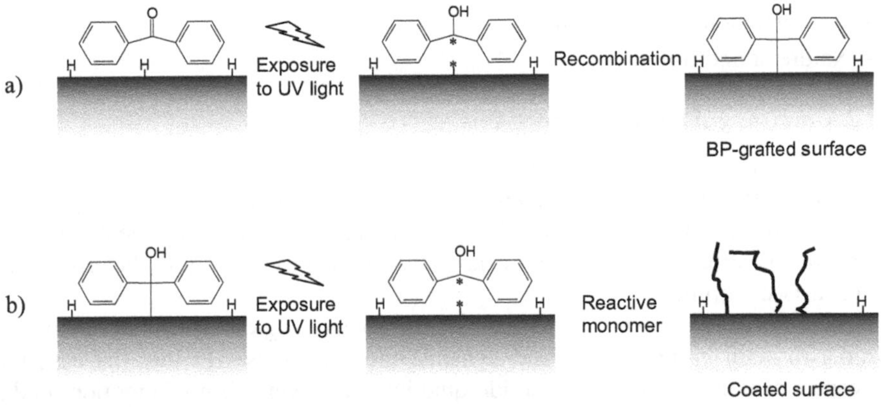

a)

b)

Figure 1. Modification of polypropylene surfaces using a two-step grafting method.

of the substrate to obtain a constant contact area in all the samples studied. The experiments were performed using the ASTM-D 4541 standard [33]. The sample was mounted in a home-built sample holder and the stud fixed to a flexible wire to ensure good alignment. The conditions used in the experiments were a speed of 1 mm/min and a maximum force of 450 N (95% of the maximum value of the load cell used). A pre-load force of 1 N was used in all the experiments. The force and displacement were monitored during the pull-off experiment. The maximum force required to separate the coating corresponded to the adhesion of the coating to the substrate. Measurements were carried out on at least 25 samples to obtain the average value and its standard deviation.

3. RESULTS AND DISCUSSION

3.1. Surface modification

The irradiation of a benzophenone type initiator excites it to a triplet state. This excited molecule can abstract a proton from the PP surface in the absence of another source and generate a radical on the surface [30]. These radicals can recombine with the semipinacol radicals, as explained in Fig. 1a. Stable species are formed with a covalent bond between the initiator and the PP surface.

These grafted molecules are called 'surface initiators'. These grafted initiators can generate radicals on exposure to UV light, the covalent bond can be broken by the effect of the UV light, and the radicals on the surface are regenerated as shown in the same figure (Fig. 1b). A series of experiments were performed in order to evaluate the different factors affecting the photo-initiator grafting reaction:

– Influence of the UV irradiation on surface properties of the PP substrate.
– Influence of the solvent for the photo-initiator on the grafting onto the PP surface.

- Concentration of photo-initiator.
- Nature of photo-initiator.

3.2. Covalent grafting of benzophenone

3.2.1. Contact angle measurements

One of the most sensitive techniques used to analyse the outermost layer of polymer surfaces is the measurement of the contact angle. Such experiments provide information about the properties of the polymer such as wettability, heterogeneity and surface mobility.

There are different methods to evaluate the contact angles. They can be divided into two main groups; static and dynamic measurements. In the first case the measurement takes place on the solid/liquid interface, which is not in motion. In the second case the liquid is in motion with respect to the solid phase. In our studies we have used the static measurement of the contact angles of water and diiodomethane to calculate the surface energy as described in the Experimental Section. The surface energy of the untreated PP was determined to be approximately 26.5 mJ/m^2 with differences between the different batches of PP. In this study, the surface energy of our PP samples is lower than the values reported in the literature. In our study we used an injection moulded sheet of a commercial grade PP whose composition is not known exactly. There are several possibilities for this low surface energy. Surface enrichment by low molecular weight fractions as a result of molecular weight distribution and composition distribution of the polyolefin might be one of the reasons; on the other hand, it might also be possible to have some release agent on the surface of the PP sheets. The effect of the irradiation on pure PP samples was almost negligible and no significant differences in the surface energy were observed during irradiation (see Fig. 2).

The use of different solvents for the grafting procedure, i.e. acetone, dichloro-methane and toluene, has no effect on the surface energy of the PP. Similar values were obtained for all the solvents tested and toluene was chosen as the solvent since it dissolves all the photo-initiators used in this study.

To study the effect of surface coverage 0.5 to 25 wt% solutions of benzophenone (BP) in toluene were prepared, and a film of 30 μm thickness was applied onto the PP and irradiated for different exposure times ranging from 30 to 600 s. No significant differences in the surface energy can be observed when 0.5 wt% of BP is used; the surface energy of the PP remains almost constant indicating that no measurable changes took place on the surface. However, a significant increase in the value of the surface energy as well as a decrease in the contact angle of water can be observed at higher concentrations of BP in the irradiated formulation (Fig. 3). The most significant increase in the surface energy is produced during the first 60 s of irradiation and after that the value remains almost constant for concentrations of BP below 5 wt%. When more concentrated BP solutions are used for the surface modification of the PP, the surface energy increases with increasing irradiation time, but to a lesser extent than in the first 60 seconds of irradiation (Fig. 4).

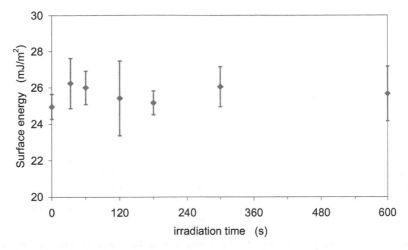

Figure 2. Variation of surface energy of the pure PP substrate as a function of UV irradiation time.

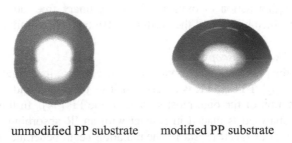

unmodified PP substrate   modified PP substrate

Figure 3. Contact angle of water on (a) an unmodified PP substrate, and (b) a photo-initiator modified PP substrate.

The increase from approx. 26 up to 34 mJ/m^2 for the photo-initiator modified PP is a clear indication that some changes had occurred at the surface. Further analysis of the surface is required to prove the presence of the BP. The contact angle of water decreases from 103° to approx. 80°; this is already an indication of the presence of hydrophilic groups on the surface originating from the grafted initiator.

The presence of oxygen during the grafting reaction has an inhibiting effect. This conclusion was drawn from surface modification experiments in the air, since no differences in the surface energy were observable after irradiation in the air even for concentrated BP solutions. In a nitrogen atmosphere a significant increase in surface energy was observed that can be associated with the grafting of the photo-initiator, as will be shown later. Different authors have observed the inhibition effect of oxygen on the grafting reaction of benzophenone; Yang and Rånby [20] concluded

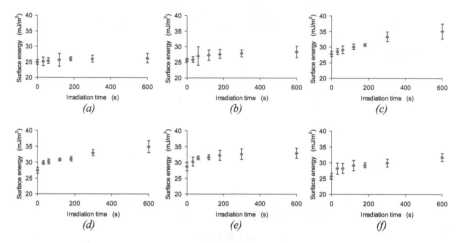

Figure 4. Surface energy of the PP vs. UV irradiation time for different concentrations of BP photo-initiator. a) no BP, b) 0.5 wt% BP, c) 1 wt% BP, d) 2 wt% BP, e) 5 wt% BP, f) 10 wt% BP.

that the scavenging action of oxygen with the primary free radicals formed was responsible for the inhibition of the grafting of BP type photo-initiators.

3.2.2. FTIR-ATR spectroscopy

The pure and the photo-initiator modified PP sheets were investigated using FTIR-ATR spectroscopy. FTIR-ATR is a very sensitive technique and allows the study of the modification of the outermost surface of the PP [34]. In this technique the surface to be analysed is placed in contact with an IR absorbing material with a high refractive index (ATR crystal). The radiation beam penetrates the sample and a part of it is reflected. The spectrophotometer measures the attenuated radiation as a function of the wavelength, obtaining the spectral characteristic for the sample. Depending on the experimental set-up (the angle of incidence of the beam) the analysed depth is different (from 4 Å to 20 micrometers) allowing the study of surfaces and their composition.

The resulting spectra of the pure polypropylene showed characteristic bands at 2953, 2916, 2872, 2839, 1452, 1375 and 1358 cm^{-1}, associated with the stretching and bending of methylene and methyne groups. The band at 1167 cm^{-1} is associated with the isotactic groups as well as the bands present at 997 and 972 cm^{-1}. They are used to quantify tacticity [35]. All these bands showed a high isotacticity of the PP used in this study (Fig. 5).

There are a few differences observable in the spectra obtained from photo-initiator modified PP surfaces when compared to the 'virgin' PP. These differences are:

1) a broad band centred at 3400 cm^{-1} that may be attributed to hydroxyl groups from the semipinacol radicals and/or hydrogen bonded water to these hydroxyl groups.

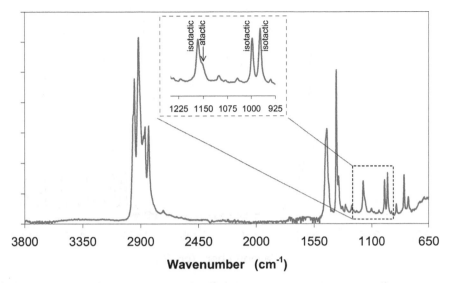

Figure 5. FTIR spectrum of the polypropylene substrate as used in this study.

2) a band at 1640 cm^{-1} that may come from the hydroxyl group of the semipinacol radical.

3) a band at 710 cm^{-1} that can be attributed to the bending of the aromatic ring.

We measured the spectra of the photo-initiator modified PP surfaces modified with a 5 wt% benzophenone formulation irradiated from 30 to 600 s. The integration of the absorption band at 710 cm^{-1} in comparison to the reference band corresponding to the bending vibration of the methyl group (1375 cm^{-1}), yielded the results shown in Fig. 6.

The increase in peak intensity at 710 cm^{-1} with increasing irradiation time indicates an increase in aromatic ring concentration. This proves again the grafting of the initiator onto the PP surface. For the other areas of interest the results are comparable, as shown in Fig. 7.

These results correspond with the ones obtained by the surface energy measurements; the surface energy increases as the irradiation time increases and by FTIR-ATR one observes a higher content of phenyl rings. The signals in the FTIR spectra are an indication that the initiator is covalently grafted onto the PP surface.

3.2.3. UV spectroscopy

For these absorption experiments, transparent PP films were used as substrates. These substrates were grafted with the photo-initiator using a 5 wt% solution of BP in toluene as described in the Experimental Section. The irradiated polypropylene has no absorption bands in the UV region nor does the non-irradiated sample. However, in the photo-initiator modified samples an absorption band around 250 nm

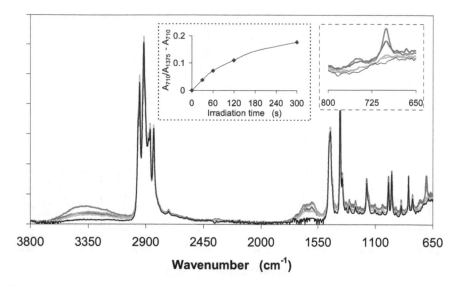

Wavenumber (cm⁻¹)

Figure 6. FTIR spectra of photo-initiator modified PP surfaces at different irradiation times, modified using a 5 wt% solution of benzophenone. Inset shows the increase of the aromatic ring vibration relative to an internal standard.

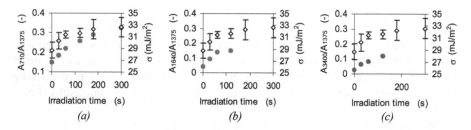

Figure 7. Comparison of the experimentally determined surface energy σ ($\diamond$) of the photo-initiator modified polypropylene and relative absorption ($\bullet$) at different UV irradiation times using a 5 wt% solution of BP.

and a higher intensity band centred at 220 nm are present (Fig. 8). These two absorption bands could be related to the grafted initiator and the concentration increases with the irradiation time until a maximum value after 300 s of irradiation is reached. The presence of this absorption in the UV region even after washing the photo-initiator modified PP substrates with abundant solvent is a clear evidence that the BP was covalently grafted onto the PP surface.

Although we have attempted to use other techniques such as DRIFT, XPS and AFM to prove photo-initiator grafting and to determine the thickness of the grafted layer, no clear conclusions could be drawn from the results. In all cases some

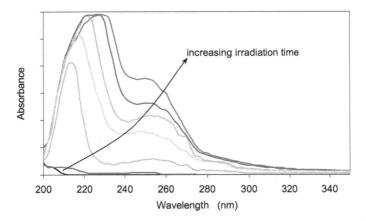

Figure 8. UV absorption spectra of an untreated PP film, and photo-initator modified PP films that were irradiated for 0 s, 30 s, 60 s, 120 s, 180 s, 240 s, and 300 s.

differences were observed for the photo-initiator modified samples, higher oxygen content and some characteristic signals originating from groups corresponding to the photo-initiator were observed; but in some cases the high noise and in other cases only small differences made a correct quantification very difficult.

3.2.4. Time stability of the grafted BP onto the PP substrate
One of the main problems of industrial methods to increase the polarity of the PP surface is the stability of the modified surface. For plasma treatments it is well known that the modified surface of the polymer is stable only for a limited period of time and after that the surface loses its hydrophilic nature [36]. To be able to maintain the surface properties of the photo-initiator modified PP for long periods of time is important from an application point of view. The supplier could do the modification of the PP and the final user could buy the modified PP and use it when needed.

In order to study the time stability of the modification, different samples of PP were modified with a solution of 5 wt% of BP irradiated for 600 s and stored in the dark. Their surface energy was monitored for a period of 3 weeks. The surface energy of the samples was found to be almost constant at about 37 mJ/m^2, during the first few days after modification and after 3 weeks the value dropped only to 35 mJ/m^2 (Fig. 9). The surface energy of our unmodified PP surfaces was approx. 26.5 mJ/m^2 so the changes observed are significant. In this study, the surface energy of our PP surfaces is lower than the values reported in the literature. In our study we used an injection moulded sheet of a commercial PP grade whose composition is not known exactly (and is not important for our study). Reasons for this low surface energy might be a surface enrichment by low molecular weight fractions as a result of molecular weight distribution and composition distribution of the polyolefin. On

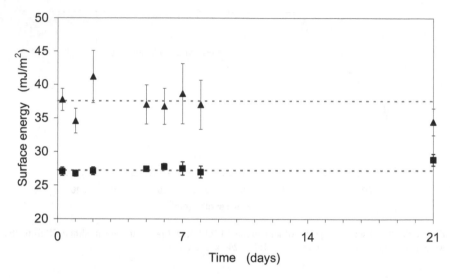

Figure 9. Surface energy vs. time of unmodified PP (■) and photo-initiator modified PP (▲) sheets using a 5 wt% BP solution.

the other hand, it might also be possible to have some release agent on the surface of the PP sheets. The grafting of photo-initiators onto our PP substrates results in a stable modification of their surface properties. Modified samples can be stored and used with the same effectiveness even after 3 weeks after their preparation.

3.3. Proof of reactivity of the photo-initiator modified PP substrates

In order to further prove the presence of the grafted photo-initiator, and the remaining activity of the grafted photo-initiator on the modified PP substrate we used the photo-initiator modified samples as substrates to graft some acrylic monomers. The photo-initiator present on the surface can generate radicals upon UV irradiation and initiate the polymerisation, as we discussed before. As a result, acrylate modified surfaces will be obtained.

The acrylic monomers chosen were acrylic acid (AA), 2-hydroxyethyl acrylate (HEA) and pentaerythritol triacrylate (PETA).

In order to check the grafting reaction the acrylates were grafted onto pure PP and on photo-initiator modified PP. Also an acrylate containing 1 wt% of benzophenone was grafted onto pure PP as control. For the irradiation experiments the surface was covered with a quartz plate of the same size as the polypropylene sheet to ensure good contact of the pure PP with the different acrylates. The irradiated samples were washed adequately with an appropriate solvent, extracted for at least 4 hours to remove the unreacted acrylate from the surface, and dried overnight.

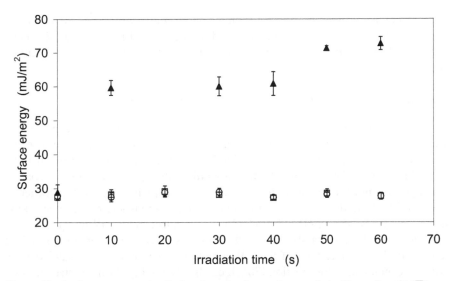

Figure 10. Surface energy vs. irradiation time of polypropylene grafted with acrylic acid (□) pure PP, (○) pure PP (AA + 1 wt% BP) and (▲) photo-initiator modified PP.

The surface energy of all samples was determined using the contact angles of water and diiodomethane, as described before.

The surface energy of the pure PP samples remains almost constant around 26.5 mJ/m^2 with irradiation time but in the case of the modified samples a large increase is observed even in the first 20 seconds of irradiation. The surface energy increases from 30 to 70 mJ/m^2 as the contact angle of water decreased from 100° to 20° in the case of AA (Fig. 10). The sample containing AA and BP as photo-initiator in the formulation that was applied onto untreated PP has a low graft efficiency as in the case of pure AA on untreated PP. It can be attributed to the different affinity of the monomer to the generated radicals, the surface radicals, the macromolecular radicals and the semipinacol radicals, which have increasing polarity in this order. The acrylic acid is a polar monomer and therefore will have a higher affinity with the semipinacol radicals initiating the polymerisation leading to homopolymerisation and a low grafting of polymeric chains. In our study we observed that the acrylates react only with the photo-initiator modified samples due to the presence of the grafted photo-initiator at the surface reaching surface energies of 70 mJ/m^2 for AA, 58 mJ/m^2 in the case of HEA and 64 mJ/m^2 when the PETA was used.

The increase in surface energy can be attributed to the grafting of the acrylates onto the photo-initiator modified surfaces due to the higher content of polar groups. The amount of grafted acrylate chains onto polypropylene increases with irradiation time and the surface energy values reach maximum values approximately after 60 s of irradiation. The fact that the acrylates do not graft onto pure PP samples confirms the effect of the grafted photo-initiator on the photo-initiator modified PP surface as expected. The grafted acrylate chains could not be washed from the surface using organic solvents indicating covalent bonding between the photo-initiator modified surface and the acrylate.

3.4. Powder coating on photo-initator modified PP substrates

As described in the Experimental Section, a mixture of two unsaturated polyesters was used as the powder coating binder. The effects of temperature, and the type and concentration of photo-initiator on the kinetics of curing were studied previously by a combination of UV rheology, DSC and FTIR studies [37]. The powder coating was applied onto the PP using a fluidised bed as explained in the Experimental Section. The conditions of curing used in this study were isothermal cure at 100°C using 2.5 wt% IC651 for 60 s of irradiation under nitrogen (20 mW/cm^2).

To understand the wettability of the PP substrates, the development of the surface tension of the polymer melt of the UV curable powder coating was studied for a cured and an uncured formulation [38]. It can be seen in Fig. 11 the surface tension of the melt has larger temperature dependence in the uncured state than in the cured state. This is expected, as the mobility of the functional groups is much lower in the cured material. From this figure it can also be seen that surface tension of the melt increases upon curing.

One of the limitations for the use of coatings on polyolefins is their poor adhesion as mentioned before. The adhesion between two materials can be measured in two ways: in terms of *forces*, defining the adhesion as the maximum force per unit area exerted to separate the two materials, or in terms of *work or energy*, defining the adhesion as the work done in separating or detaching the two materials.

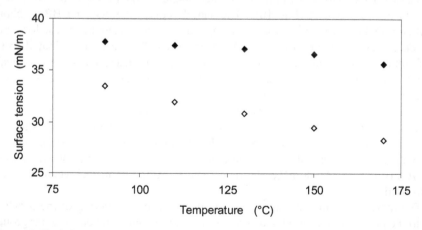

Figure 11. Evolution of the surface tension of the powder coating formulation, as applied onto the PP substrate, as a function of temperature (◇) uncured coating formulation, (◆) cured coating formulation.

A large number of techniques for adhesion measurement of coatings have been described in the literature [39–41]; these techniques range from inexpensive to very sumptuous and from primitive to very sophisticated. All measurement methods measure practical adhesion. There is no single technique, which will be acceptable to everyone or will be applicable to all coating-substrate combinations. For relative purposes any of these techniques can be used and will be able to rank or discriminate cases of poor adhesion. It is widely realised that an adhesion test method must be selected on the basis of the nature of the coating-substrate combination, application and so on. No universal test is available; different tests yield different results.

In our study we decided to use the pull-off method to measure the adhesion. In this method a loading fixture, commonly called a dolly or a stud, is glued to the surface of the coating. A special device is then used to apply an increasing force until the coating or the glue fails. The pull direction is perpendicular to the surface, so the tensile strength is measured.

Coatings were applied on differently pre-treated PP substrates. The effect of grafting of the photo-initiator and/or sanding was evaluated. Figure 12 shows the pull-off force required to detach the coating from the different substrates.

It can be clearly seen that the adhesion of the coating to pure PP is very low, a force of approximately 20 N (0.38 MPa) is required to detach the coating. In all these experiments a failure between the coating and the substrate is observed. This was verified by surface energy measurements on the fractured surfaces of the coating and the substrate after delamination; the surface energy of the PP after delamination is approximately equal to the surface energy of the PP as received,

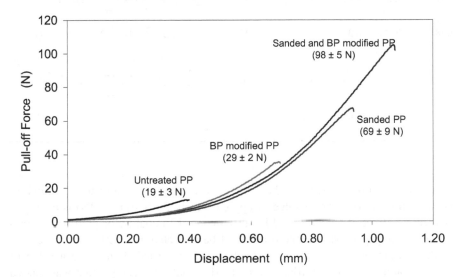

Figure 12. Pull-off force measured to detach the cured coatings from different substrates.

Table 1.
Surface energy and its components (in mJ/m^2) and adhesion performance in terms of force required to detach the powder coating from the photo-initiator modified PP substrates

	Surface energy	Dispersion component	Polar component	Dispersion/ Total	Structure of the BP derivative	Pull-off force (N)
Uncured coating	40.61	40.12	0.49	0.988		
Cured coating	41.34	40.72	0.62	0.985		
Untreated PP	25.2	24.45	0.75	0.970		19
Ini1	33.87	33.1	0.77	0.977	*[benzophenone structure]*	98
Ini2	32.43	30.7	1.73	0.947	CH$_3$(CH$_2$)$_7$O— *[BP derivative structure, OH]*	62.1
Ini3	31.02	30.24	0.78	0.975	HO— *[BP derivative structure, OH]*	100.6
Ini4	32.51	30.87	1.64	0.950	HO— *[BP derivative]* —OH	76.7
Ini6	31.02	30.12	0.9	0.971	CH$_3$(CH$_2$)$_7$O / CH$_3$(CH$_2$)$_7$O— *[BP derivative]*	93.1
Ini7	33.27	33.02	0.25	0.992	CH$_3$(CH$_2$)$_7$O— *[BP derivative]* —O(CH$_2$)$_7$CH$_3$	69.2
Ini8	30.42	30.3	0.12	0.996	CH$_3$(CH$_2$)$_7$O— *[BP derivative]*	71.1
Ini9	24.18	23.67	0.51	0.979	CH$_3$(CH$_2$)$_7$O / CH$_3$(CH$_2$)$_7$O / CH$_3$(CH$_2$)$_7$O— *[BP derivative]*	71.6
Ini10	36.71	35.85	0.86	0.977	*[alkyl-BP derivative structure]*	59.9

as well as the surface energy of the cured coating. The photo-initiator modified PP presents higher adhesion; a value of approximately 30 N (0.60 MPa) is obtained. A positive effect of sanding the PP substrate prior to coating or grafting, using a 400-grit sandpaper, was observed as could be expected as a result of an increase in the roughness of the substrate. In the case of a sanded substrate a value of approximately 70 N (1.40 MPa) is measured. Finally the sample that was sanded and afterwards modified with the photo-initator procedure described in this paper showed the highest adhesion. A force of approximately 100 N (2.00 MPa) is obtained. This value of adhesion is in the range of an acceptable level for adhesion of coatings onto plastic substrates.

When performing a crosscut adhesion test on these samples [42], a value of 5 for both the pure PP and the photo-initiator modified PP was observed. The sanded samples showed a higher adhesion, the sanded PP had a rating of 4 while the sanded and photo-initiator modified PP had a rating of 1 which means an acceptable adhesion. It has to be stated here that the crosscut adhesion test is a rather difficult test to perform on polymeric substrates.

3.5. Effect of modification of benzophenone moiety on adhesion results

The effect of the structure (or polarity) of the surface grafted BP derivative on the adhesion was also tested in this research. The thickness of the applied powder coating (containing 2.5 wt% IC651 and cured at 100°C in a nitrogen atmosphere) was approx. 100 μm. For the powder coating formulation as used, the surface energy of the uncured as well as the cured coating was determined, and the results are shown in Table 1.

All of the modified BP initiators that were grafted onto the PP substrate showed a significant increase of the adhesion of the powder coating with respect to the untreated PP. When comparing the types of modification of the BP moiety, the overall increase in pull-off force and the value of the surface energy of the photo-initiator modified PP surface are both affected. The results are summarised in Table 1.

Based on these results it can be concluded that the requirements for good adhesion of the UV curable powder coating formulation is a surface energy of the substrate of approx. 32 mJ/m^2 with a ratio of dispersion component/total surface energy of approx. 0.98 to obtain good adhesion (i.e. forces of at least 90 N are needed for failure).

4. CONCLUSIONS

The grafting of benzophenone type photo-initiators onto polypropylene substrates was achieved using UV light. As soon as the UV light hits the sample and the concentration of initiator is beyond 2 wt% a covalent bonding of the photo-initiator to the surface is observed. Contact angle measurements, FTIR-ATR and UV spectroscopy confirmed the presence of the initiator on the surface, and the concentration of grafted moieties increases with increasing irradiation time. In most of the cases it reaches a 'surface saturation plateau' after approximately 60 s of irradiation, where the maximum amount of initiator is grafted onto the surface. Longer irradiation times seemed to have no further effect on the surface energy values. The activity of the surface grafted photo-initiator remains intact during storage; photo-initiator modified samples can be stored and used with the same effectiveness even after 3 weeks after their preparation.

Grafting of different monomeric acrylates on the photo-initiator modified polypropylene samples as substrates was possible since the photo-initiator on the surface acts as an initiation centre leading to grafted chains on the surface with covalent bonding between the growing chains and the surface itself. The amount of grafted chains increases with increasing irradiation time and reaches a maximal value around 60 s.

UV curable powder coatings were applied efficiently onto polypropylene substrates and cured with UV light at relatively low temperatures. The adhesion of the coatings was measured with a pull-off test. Coatings applied on pure PP showed a very poor adhesion. The grafting of benzophenone derivatives on the PP increases the adhesion of the coating. Finally sanding the PP and grafting benzophenone derivatives onto it prior to the application of the coating leads to an acceptable value of adhesion of approximately 2.0 MPa.

The use of this approach is versatile and it can be used for different types of polymer substrates and different types of UV curable coating formulations. This research is performed on a lab-scale but it can be imagined that it is also applicable on a more industrial level where the photo-initiator modification step (e.g. spray application) and UV irradiation are in the same processing line as the coating step (with UV irradiation).

REFERENCES

1. K. L. Mittal (Ed.), *Polymer Surface Modification: Relevance to Adhesion*. Vol. 4, VSP/Brill, Leiden (2007).
2. K. L. Mittal (Ed.), *Polymer Surface Modification: Relevance to Adhesion*. Vol. 3, VSP/Brill, Leiden (2004).
3. N. Steward and M. J. Walzak, *Mod. Plast. Int.* **23**, 69 (1995).
4. R. A. Ryntz, *Prog. Org. Coat.* **25**, 73 (1994).
5. F. Garbassi, M. Morra and E. Occhiello, *Polymer Surfaces*, Wiley, New York (1994).
6. C.-M. Chan, *Polymer Surface Modification and Characterization*, Hanser, New York (1994).
7. L. H. Lee, *J. Adhesion Sci. Technol.* **5**, 71 (1991).
8. H. Hintze-Bruning and H. Borgholte, *Prog. Org. Coat.* **40**, 49 (2000).
9. F. Severini, L. Di Landro, L. Galfetti, L. Meda, G. Ricca and G. Zenere, *Macromol. Symp.* **181**, 225 (2002).
10. C. Muhlan and H. Nowack, *Surf. Coat. Tech.* **98**, 1107 (1998).
11. M. S. Kang, B. Chun and S. S. Kim, *J. Appl. Polym. Sci.* **81**, 1555 (2001).
12. C. Y. Huang and C.-L. Chen, *Surf. Coat. Tech.* **153**, 194 (2002).
13. F. Denes, *Trend. Polym. Sci.* **5**, 23 (1997).
14. M. Ratzsch, H. Bucka and C. Wohlfahrt, *Angew. Makromol. Chem.* **229**, 145 (1995).
15. F. Picchioni, J. G. P. Goossens and M. van Duin, *Macromol. Symp.* **176**, 245 (2001).
16. M. Ratzsch, M. Arnold, E. Borsig, H. Bucka and N. Reichelt, *Prog. Polym. Sci.* **27**, 1195 (2002).
17. K. Allmer, A. Hult and B. Rånby, *J. Polym. Sci.: Part A: Polym. Chem.* **26**, 2099 (1988).
18. Y. Uyama and Y. Ikada, *J. Appl. Polym. Sci.* **36**, 1087 (1988).
19. W. Beenen, D. J. Wal and L. P. B. M. Janssen, *Macromol. Symp.* **102**, 255 (1996).
20. W. Yang and B. Rånby, *J. Appl. Polym. Sci.* **62**, 533 (1996).
21. W. Yang and B. Rånby, *J. Appl. Polym. Sci.* **63**, 1723 (1997).

22. Y. Li, J. M. DeSimone, C.-D. Poon and E. T. Samulski, *J. Appl. Polym. Sci.* **64**, 883 (1997).
23. C. Decker and K. Zahouily, *Macromol. Symp.* **129**, 99 (1998).
24. C.-W. Lin and W.-L. Lee, *J. Appl. Polym. Sci.* **70**, 383 (1998).
25. H. J. Chun, S. M. Cho, Y. M. Lee, H. K. Lee, T. S. Suh and K. S. Shinn, *J. Appl. Polym. Sci.* **72**, 251 (1999).
26. S. G. Flores-Gallardo, S. Sanchez-Valdes and L. F. Ramos de Valle, *J. Appl. Polym. Sci.* **79**, 1497 (2001).
27. E. Borsig, M. Lazar, A. Fiedlerova, L. Hrckova, M. Ratzsch and A. Marcincin, *Macromol. Symp.* **176**, 289 (2001).
28. G. Oster, G. K. Oster and H. J. Moroson, *J. Polym. Sci.* **XXXIV**, 671 (1959).
29. C. Decker, *J. Appl. Polym. Sci.* **28**, 97 (1983).
30. H. Ma, R. H. Davis and C. N. Bowman, *Macromolecules* **33**, 331 (2000).
31. P. Castell, M. Wouters, G. de With, H. Fischer and F. Huijs, *J. Appl. Polym. Sci.* **92**, 2341 (2004).
32. D. K. Owens and R. C. Wendt, *J. Appl. Polym. Sci.* **13**, 1741 (1969).
33. ASTM D 4541 'Standard Test Method for Pull-Off Strength of Coatings using Portable Adhesion Testers' (2002).
34. N. S. Harrick, *Internal Reflection Spectroscopy*, Wiley, New York (1967).
35. J. P. Luongo, *J. Appl. Polym. Sci.* **3**, 302 (1960).
36. H. Yasuda, *Plasma Polymerization*, Academic Press, New York (1985).
37. P. Castell, M. Wouters, G. de With and H. Fischer, submitted.
38. M. Wouters and B. de Ruiter, *Prog. Org. Coat.* **48**, 207 (2003).
39. K. L. Mittal, *Polym. Eng. Sci.* **17**, 467 (1977).
40. K. L. Mittal, in *Adhesion Measurement of Thin Films, Thick Films and Bulk Coatings*, K. L. Mittal (Ed.), pp. 5-17, STP640, ASTM, Philadelphia (1978).
41. K. L. Mittal, in *Adhesion Measurement of Films and Coatings*, K. L. Mittal (Ed.), pp. 1-13, VSP/Brill, Leiden (1995).
42. ASTM D 3359 'Standard Test Methods for Measuring Adhesion by Tape Test' (2002).

Contact Angle, Wettability and Adhesion, Vol. 5, pp. 229–238
Ed. K.L. Mittal
© VSP 2008

Modulation of the surface properties of reactive polymers by photo-Fries rearrangement

SUSANNE TEMMEL,[1] THOMAS HÖFLER[2] and WOLFGANG KERN[2,3,*]

[1]*Polymer Competence Center Leoben GmbH, Parkstrasse 11, A-8700 Leoben, Austria*
[2]*Graz University of Technology, Institute for Chemistry and Technology of Organic Materials, A-8010 Graz, Austria*
[3]*Montanuniversität Leoben, Institute of Chemistry of Polymeric Materials, A-8700 Leoben, Austria*

Abstract—The UV irradiation of polymers containing phenyl ester units yields o- and p-hydroxyphenone groups (photo-Fries rearrangement). In this paper we report on the photo-induced changes in the surface properties of poly(4-acetoxystyrene) as a model polymer. The changes in surface energy and streaming potential were studied as a function of UV illumination time. Both contact angle and zeta potential measurements confirmed that acidic phenolic units were generated at the polymer surface upon UV illumination. These findings were supported by FTIR data. Photoreactive polymers of this type are of interest as functional surface coatings for a variety of substrates.

Keywords: Photochemistry; photo-Fries rearrangement; poly(4-acetoxystyrene); FTIR; contact angle; zeta potential.

1. INTRODUCTION

The well-known photo-Fries rearrangement provides a convenient and attractive way for the synthesis of o- and p-hydroxyaromatic ketones. In this photoreaction the acyl group of an aryl ester shifts to the ortho or para position of the aromatic ring. As a result, an aromatic keton bearing a phenolic OH unit is generated. This rearrangement was first described by Anderson and Reese in 1960 [1]. The reaction has been studied by many authors [2–12]. The generally accepted mechanism [13] implies as the first step the homolytic cleavage of the C-O band which proceeds from an excited singlet (S_1) state. For phenyl acetate as an example, an acetyl and a phenoxy radical are generated within the solvent cage. Recombination of this geminate radical pair (within the cage) then yields a cyclodienone structure which is transformed to the tautomeric hydroxyketone. The ratio of the ortho- and para-substituted products can be controlled by the reaction temperature. Out-of-cage

---

*To whom correspondence should be addressed. Tel.: ++43 3842 402 2350;
Fax: ++43 3842 402 2352; e-mail: wolfgang.kern@unileoben.ac.at

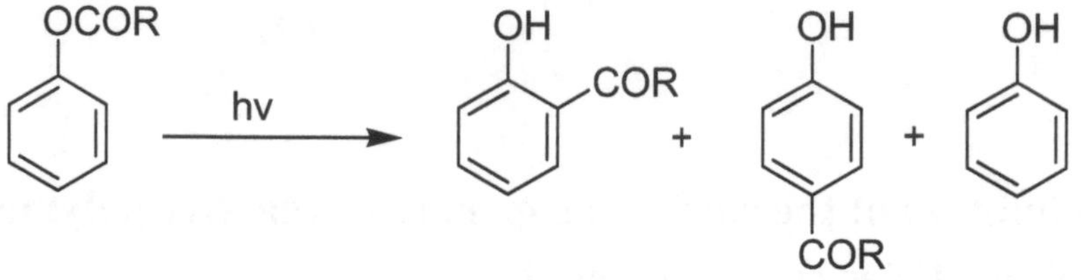

Scheme 1. General reaction of an aryl ester, which yields a mixture of ortho- and para-acylphenols and phenol as a by-product.

reaction of the aryloxy radical was postulated to account for phenol formation by hydrogen abstraction from the solvent. The overall reaction of an O-aryl ester is represented in Scheme 1. In addition to this, recent studies have shown that in aryl esters decarboxylation and decarbonylation reactions can occur as a side reaction [14].

The aim of the present study was a detailed investigation of the surface properties of photosensitive polymers containing O-aryl groups. We have chosen poly(4-acetoxystyrene) as a model polymer for these investigations. This polymer is known as a positive photoresist material [15]. In a recent publication [16] we have shown that the rearrangement of the acyl group results in a large increase of the refractive index in this polymer and in similar materials. While this effect can be utilized for the lithographic inscription of waveguides and holographic data storage, the present investigation focuses on the changes in the surface properties of poly(4-acetoxystyrene) that can be achieved by UV illumination. The irradiated surfaces are promising for a further surface functionalization by selected post-exposure reactions.

2. EXPERIMENTAL SECTION

2.1. Preparation of poly(4-acetoxystyrene)

Poly(4-acetoxystyrene) was prepared by a free-radical polymerization of 4-acetoxystyrene using 2,2′-azo-bis(2-methylpropionitrile) (α,α'-azoisobutyronitrile, AIBN) as the initiator. The polymerization process is represented in Scheme 2. The reaction was carried out under anhydrous and inert atmosphere conditions. A mixture of 4-acetoxystyrene (10 mL, 0.09 mol; from Sigma Aldrich) and AIBN (100 mg, 0.61 mmol; from Fluka) was dissolved in anhydrous toluene (100 mL) and heated to 100°C under nitrogen flow. After five hours the polymerization was stopped by rapid cooling. The polymer was then precipitated by dropping the solution slowly and under effective stirring into 500 mL of methanol. The precipitate was collected by filtering the mixture and thoroughly washed with methanol. The polymer was dried overnight *in vacuo* (2 kPa) at 40°C until the weight became constant.

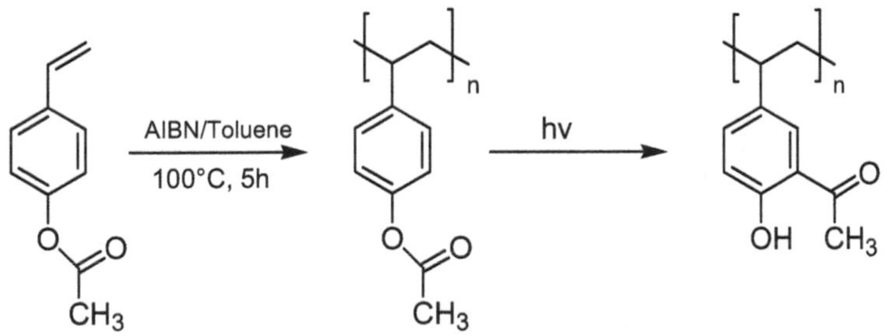

Scheme 2. Free radical polymerization of 4-acetoxystyrene and photo-Fries reaction of poly(4-acetoxystyrene).

An overall yield of 30% of poly(4-acetoxystyrene) was achieved. The molecular weight distribution of poly(4-acetoxystyrene) was determined by gel permeation chromatography (GPC) using $CHCl_3$ as eluent (calibration against polystyrene standards) and $M_w = 19000$ g mol^{-1} and $M_n = 14000$ g mol^{-1} were found. FTIR spectra were taken with a Perkin Elmer instrument (Spectrum One). In this case, films of the polymer were cast onto NaCl plates. For poly(4-acetoxystyrene) the bands at 1763 cm^{-1} (C=O of aromatic esters) and 1198 cm^{-1} (C-O-C of the ester group) are typical. UV/Vis spectra were recorded with a JASCO V-530 spectrophotometer. The UV absorption of films of poly(4-acetoxystyrene) (cast onto CaF_2 plates) extends to 290 nm which is typical of the phenyl chromophore. From Fabry-Perot interference fringes in the UV/Vis range the thickness of the cast polymer films was estimated.

2.2. Sample preparation and UV irradiation

Films of poly(4-acetoxystyrene) were prepared on silicon wafers. To improve the adhesion of the polymer film to silicon, the wafers were primed with spin-on glass (Honeywell Electronic Materials, Califorina, USA) by spin-casting (3000 rpm, 45 s). The wafer was then baked at 80°C for 60 minutes to eliminate the excess solvent (soft bake). After this pre-treatment, a layer of the polymer (thickness between 300 and 400 nm) was spin-cast onto the wafer from $CHCl_3$ solution (10 g/L). The polymer film was dried at 40°C *in vacuo* (2 kPa) for 60 minutes. The sample was then positioned in a reaction chamber made from stainless steel, which was equipped with a quartz window and gas in/outlets. The sample was purged with nitrogen for 20 min to remove the oxygen and then UV irradiated for different periods of time (5 seconds to 10 minutes) while maintaining the nitrogen gas flow (10 L/hr). After UV illumination, the chamber was purged with nitrogen for additional 5 minutes.

All UV irradiations were carried out with a medium pressure mercury lamp (Heraeus, 1300 W). An etched (i.e. corrugated) quartz plate was positioned as a diffuser element between the lamp and the reaction chamber in order to provide a

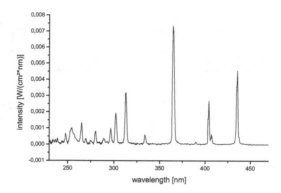

Figure 1. Emission spectrum of the Hg-lamp as recorded with the spectroradiometer. The intensity (i.e. power density) was measured in the sample plane.

uniform light intensity over the whole sample area. The distance between the lamp and the sample surface was 7.2 cm. The UV radiation impinging on the sample surface was measured with a UV spectroradiometer from Solatell Ltd. (Croydon, UK) with a measuring range from 230 nm to 465 nm. The spectral distribution of the radiation is represented in Fig. 1. The power density I in the sample plane was $I = 72$ mW/cm^2 (integrated over the spectral range 230–465 nm) and $I = 17$ mW/cm^2 (integrated over the spectral range 230–290 nm), respectively.

2.3. Contact angle measurements

The surface energy γ of the sample surfaces was determined by measuring contact angles θ with a Drop Shape Analysis System DSA100 (Krüss GmbH, Hamburg, Germany) using water and diiodomethane as test liquids (drop volume $\sim 4~\mu$L). Based on the Owens-Wendt method [17], the surface energy γ as well as the dispersion and polar components (γ^D and γ^P) were determined. The literature values for the surface tension components of the test liquids used are given in Table 1 [18]. The contact angles were obtained using the sessile drop method and were measured within two seconds after deposition of the droplet. The reproducibility

Table 1.
Surface tension γ, dispersion component γ^D, polar component γ^P, density and viscosity values for water and diiodomethane. Data taken from Ström *et al.* [18]

	Temperature [°C]	γ [mN/m]	γ^D [mN/m]	γ^P [mN/m]	Density [g/cm^3]	Viscosity η [mPa*s]
Water	25	72.8	21.8	51	0.998	1.002
Diiodomethane	20	50.8	50.8	0	3.325	2.762

was within 2°. All measurements were performed at room temperature. The data given are average values over four individual measurements.

2.4. Zeta potential measurements

The zeta potential ζ of the sample surfaces was determined by the streaming potential method, using a clamping cell connected to an EKA Electrokinetic Analyzer (Anton Paar GmbH, Graz, Austria). In this cell the sample is pressed against a poly(methyl methacrylate) (PMMA) spacer with rectangular channels. The measurements were performed with a KCl electrolyte solution (10^{-3} M, 500 mL). The pH was adjusted to about 4 by adding HCl (0.1 M, 0.4 mL) and was then increased stepwise (0.3–0.4 units) by titration with NaOH (0.05 M), until pH = 9 was reached. Using the clamping cell, a pressure ramp from 0 to 40 kPa was employed to force the electrolyte solution through the cell along the rectangular channels of the poly-(methyl methacrylate) spacer. Streaming potentials were converted to zeta potentials using the Helmholtz-Smoluchowski equation [19] and the Fairbrother-Mastin approach [20]. Each value of the zeta potential at a given pH value represents an average value over at least three individual measurements.

3. RESULTS AND DISCUSSION

3.1. FTIR spectroscopy

FTIR spectroscopy was used to follow the expected photo-Fries rearrangement in poly(4-acetoxystyrene) as visualized in Scheme 2. Figure 2 gives a comparison of the FTIR spectra of films of poly(4-acetoxystyrene) prior to and after irradiation. In this case, the films (thickness between 300 and 400 nm; on NaCl plates) were exposed to polychromatic UV light under nitrogen atmosphere for 6 minutes.

In the spectrum of the non-irradiated film the signals at 1763 (C=O stretch) and at 1198 cm^{-1} (asym. C-O-C stretch) are typical of the ester units. The position of these two bands exactly meets the expectation for esters R_1-(C=O)-O-R_2 with R_1 being an aliphatic unit and R_2 being a phenyl ring [21].

Other bands in this FTIR spectrum are typical of aliphatic groups (2930 cm^{-1}: C-H stretch; 1370 cm^{-1}: C-H deformation in the CH$_3$ units of the acetate group) and aromatic groups (1604 and 1505 cm^{-1}: aromatic ring vibration, 847 cm^{-1}: C-H deformation in 1,4-disubstituted benzene). The signals at 1016 and 911 cm^{-1} are related to the acetate group.

After UV irradiation, significant changes are observed in the FTIR spectrum of poly(4-acetoxystyrene). The signals of the phenyl ester group at 1763 and 1198 cm^{-1} have almost disappeared. Also the signals at 1016 and 911 cm^{-1} have vanished while the band at 847 cm^{-1} has lost in intensity and has been shifted to 831 cm^{-1}. New bands are observable at 3400 and 1641 cm^{-1}. These signals indicate the formation of a hydroxyketone which is the expected photo-Fries product (cf.

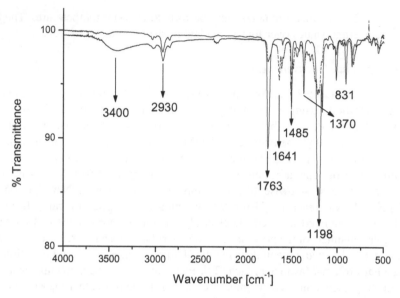

Figure 2. FTIR spectrum of poly(4-acetoxystyrene) before (—) and after UV irradiation for 6 minutes (- - -) under inert gas conditions.

Scheme 2). The broad band at 3400 cm^{-1} stems from the O-H stretching vibration of hydroxyl groups, while the split carbonyl band with peak maxima at 1641 and 1613 cm^{-1} is typical of aromatic-aliphatic ketones. For comparison, the FTIR spectrum of 2-hydroxyacetophenone, a low-molecular-weight reference compound, was recorded. For this compound a split carbonyl band at 1643 and 1617 cm^{-1} is found. These FTIR data prove that in poly(4-acetoxystyrene) the photo-Fries reaction proceeded as expected.

However, from these FTIR spectra it is difficult to assess the amount of side-products (e.g. decarboxylation and formation of free phenols). To estimate the photochemical yield of o-hydroxy ketone units (i.e. the photo-Fries product) in poly(4-acetoxystyrene), the IR absorbance coefficients of model compounds (acetoxybenzene and 2-hydroxyacetophenone) were determined. A comparison of the absorbance of the ester carbonyl peak (1763 cm^{-1}) in non-irradiated poly(4-acetoxystyrene) with the absorbance of the ketone carbonyl peak (1641 cm^{-1}) in UV illuminated poly(4-acetoxystyrene) showed that the yield of ketone is between 30 and 35% after 10 min of irradiation. This is the maximum value for this polymer; when the UV irradiation was prolonged, the intensity of the ketone peak decreased again.

3.2. Contact angle measurements

Poly(4-acetoxystyrene) surfaces were characterized by contact angle measurements before and after UV irradiation for different periods of time. As shown in Fig. 3, the water contact angle θ on the non-irradiated polymer surface is high ($\theta = 97°$), whereas the diiodomethane contact angle is very low ($\theta = 22°$). These values indicate the hydrophobic character of this polymer. After UV irradiation changes in the contact angles become apparent. While the contact angle of water decreased only slightly from $\theta = 97°$ to $\theta = 90°$ after 6 min of UV irradiation, the contact angle of diiodomethane increased from $\theta = 22°$ to $\theta = 50°$. These

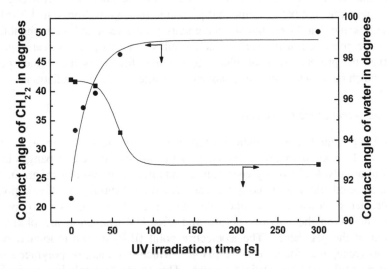

Figure 3. Contact angles θ of diiodomethane and water on poly(4-acetoxystyrene) as a function of UV irradiation time under nitrogen atmosphere. Data points represent an average over four individual measurements (the solid line is a guide for the eye).

Table 2.
Contact angle and surface energy data of poly(4-acetoxystyrene) subjected to UV irradiation. γ: surface energy; γ^D dispersion component; γ^P polar component; surface polarity = 100 (γ^P / γ)

Irradiation time [s]	θ [°] for water	θ [°] for diiodomethane	γ [mJ/m^2]	γ^D [mJ/m^2]	γ^P [mJ/m^2]	Surface polarity [%]
0	97.0	21.6	47.3	47.3	0.0	0.0
5	96.9	33.3	42.9	42.8	0.1	0.2
15	98.5	37.2	41.0	40.9	0.1	0.2
30	96.7	39.7	40.0	39.8	0.2	0.5
60	94.4	46.3	37.0	36.3	0.7	1.9
300	92.8	50.1	35.3	34.2	1.1	3.1
600	90.0	51.4	35.2	33.5	1.7	4.8

values demonstrate that the photo-Fries rearrangement made the surface of poly(4-acetoxystyrene) more hydrophilic and at the same time more oleophobic. This behavior is consistent with the formation of phenolic OH groups at the surface.

Table 2 includes surface energies γ calculated from the water and diiodomethane contact angles for the polymer surfaces. The surface energy (γ) is made up of two components, the dispersion (γ^D) and the polar (γ^P). Concerning non-irradiated poly(4-acetoxystyrene), its surface energy is almost entirely due to the dispersion component, the polar component being almost zero. This demonstrates the absence of polar groups. Upon UV illumination, the dispersion component γ^D decreased significantly from 47.3 to 33.5 mJ m^{-2} as a result of the photo-Fries rearrangement. At the same time, the polar component γ^P increased slightly from 0 to 1.7 mJ m^{-2}. It is remarkable that the overall surface energy γ decreases, although phenolic OH groups are generated at the surface as a result of the photo-Fries rearrangement. However, the presence of hydrophilic groups is indicated by the surface polarity γ^P/γ, which increases from zero to approx. 5% after 6 min of irradiation.

3.3. Zeta potential measurements

Streaming potential measurements provide information on the acid/base properties and on the hydrophilic/hydrophobic character of sample surfaces. Changes in the ζ potential of poly(4-acetoxystyrene) before and after the photo-Fries rearrangement can be detected and interpreted quite well. The reproducibility of all zeta potential measurements shown in this paper was within 2 mV. In Fig. 4 the ζ potential measurements on the polymer before and after UV irradiation are plotted as a function of the pH value. The isoelectric point (IEP) of non-irradiated poly(4-acetoxystyrene) was found to be at pH 4.3, which is typical of polymer surfaces having no acidic or basic surface groups. Due to the hydrophobic nature of the sample (see Section 3.2), a preferred adsorption of hydroxide (and also chloride anions) is observed which gives a negative zeta potential in the range between pH = 4.3 and pH = 9.0.

Compared to non-irradiated poly(4-acetoxystyrene), a shift of the isoelectric point (IEP) towards lower pH and a shift of the zeta potential ζ to more negative values is detected upon UV irradiation. At pH = 7, the zeta potential shifts from $\zeta = -28$ mV to $\zeta = -40$ mV. This is attributed to the presence of phenolic OH groups at the irradiated surface. Considering the pK$_a$ of free phenol (pK$_a \sim 10$), these OH groups confer a slightly acidic character to the polymer surface as a result of the photo-Fries rearrangement. Of course it has to be taken into account that for o-hydroxyketones the acidity of the OH group will be lowered due to the formation of hydrogen bonds. The data provide additional evidence of the photo-Fries rearrangement, which strongly influences the zeta potential of the polymer surface. As displayed in Fig. 4, the most significant changes in surface chemistry occur during the first 60 s of UV irradiation. The additional changes observed after 5 and 10 min of illumination are smaller but still noticeable.

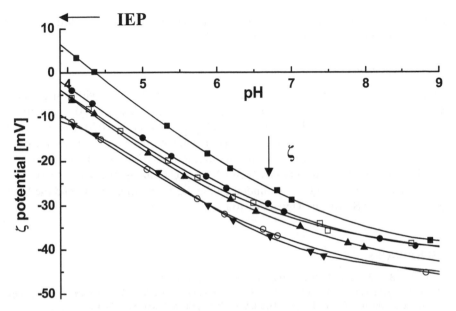

Figure 4. Zeta potential versus pH curves of poly(4-acetoxystyrene) surfaces before (■) and after UV irradiation (● 15 s, □ 30 s; ▲ 60 s, ○ 300 sec, ▼ 600 s of UV irradiation). The arrows indicate the shift of the isoelectric point (IEP) and the zeta potential (ζ).

4. CONCLUSIONS

We have demonstrated that the photo-Fries rearrangement in poly(4-acetoxystyrene) proceeds with an overall yield of 30–35%. The photoreaction can be monitored by FTIR spectroscopy which clearly shows the transformation of the aryl ester to the corresponding hydroxyketone. At the same time, the surface properties of poly(4-acetoxystyrene) change significantly. The contact angle of water decreased slightly after irradiation, whereas the contact angle of diiodomethane strongly increased. While the surface polarity increased, the overall surface energy decreased with continued UV irradiation.

These results are supported by electrokinetic investigations. It has been shown for surfaces of poly(4-acetoxystyrene) that the zeta potential shifts to more negative values upon UV illumination. Also the shift in the IEP indicates that phenolic OH groups are generated at the polymer surface. Summing up, both zeta potential and contact angle measurements are highly sensitive tools to follow the photo-Fries rearrangement at photoreactive surfaces.

The photoassisted generation of both hydroxyl and keto groups at surfaces is an interesting process which facilitates additional surface modifications. Both functional groups can be used for further transformations which is the topic of our current research [22]. Applying modern photolithographic methods, such surface modifications can be carried out in a patterned fashion at resolutions in the micrometer and sub-micrometer regimes.

Acknowledgements

Financial support by the *Austrian Science Fund* (FWF, Vienna) within the framework of the *Austrian Nano Initiative* (research project cluster 0700 – Integrated Organic Sensor and Optoelectronics Technologies, project no. 0703) is gratefully acknowledged. Part of this study was performed in the *Polymer Competence Center Leoben GmbH* (PCCL, Austria) within the framework of the Kplus-program with contributions of Graz University of Technology (TU Graz, Austria) and Anton Paar GmbH (Graz, Austria). PCCL is funded by the Austrian Government and the State Governments of Styria and Upper Austria.

REFERENCES

1. J. C. Anderson and C. B. Reese, *Proc. Chem. Soc., London*, 217 (1960).
2. M. A. Miranda and F. Galindo, in: *CRC Handbook of Organic Photochemistry and Photobiology*, 2nd ed., chap. 42, W. M. Horspool (Ed.), CRC Press, Boca Raton, FL (2004).
3. D. Bellus and P. Hrdlovic, *Chem. Rev.*, **67**, 599 (1967).
4. D. Bellus, *Adv. Photochem.*, **8**, 109 (1971).
5. H. Kobsa, *J. Org. Chem.*, **27**, 2293 (1962).
6. R. A. Finnegan and J. J. Mattice, *Tetrahedron*, **21**, 1015 (1965).
7. G. M. Coppinger and E. R. Bell, *J. Phys. Chem.*, **70**, 3479 (1966).
8. M. R. Sandner and D. J. Trecker, *J. Amer. Chem. Soc.*, **89**, 5725 (1967).
9. D. A. Plank, *Tetrahedron Lett.*, 5423 (1968).
10. J. W. Meyer and G. S. Hammond, *J. Am. Chem. Soc.*, **94**, 2219 (1972).
11. W. Adam, J. A. de Sanabia and H. Fischer, *J. Org. Chem.*, **38**, 2571 (1973).
12. C. E. Kalmus and D. M. Hercules, *J. Am. Chem. Soc.*, **96**, 449 (1974).
13. S. Lochbrunner, M. Zissler, J. Piel, E. Riedle, A. Spiegel and T. Bach, *J. Chem. Phys.*, **120**, 11634 (2004).
14. W. Gu, D. J. Abdallah and R. G. Weiss, *J. Photochem. Photobiol. A*, **139**, 79 (2001).
15. J. M. Frechet, T. G. Tessier, C. G. Wilson and H. Ito, *Macromolecules*, **18**, 317 (1985).
16. T. Höfler, T. Grießer, X. Gstrein, G. Trimmel, G. Jakopic and W. Kern, *Polymer*, **48**, 1930 (2007).
17. D. K. Owens and R. C. Wendt, *J. Appl. Polym. Sci.*, **13**, 1741 (1969).
18. G. Ström, M. Frederiksson and P. Stenius, *J. Colloid Interface Sci.*, **119** (2), 352 (1987).
19. M. Smoluchowski, *Handbook of Electricity and Magnetism*, Vol. 2, Barth, Leipzig (1921).
20. F. Fairbrother and H. Mastin, *J. Chem. Soc., Trans.*, **125**, 2319 (1924).
21. G. Socrates, *Infrared Characteristic Group Frequencies*, John Wiley and Sons Ltd., Chichester (1994).
22. T. Griesser, T. Höfler, S. Temmel, W. Kern and G. Trimmel, *Chem. Mater.* **19**, 3011 (2007).

Contact Angle, Wettability and Adhesion, Vol. 5, pp. 239–252
Ed. K.L. Mittal
© VSP 2008

Effect of grafting efficiency on peel strength, contact angle, particle size and viscosity of butyl acrylate–PUD hybrid adhesives for plastic laminates

SARTHAK K. PATEL,* R. N. JAGTAP[†] and NITIN KHATIK

University Institute of Chemical Technology, Mumbai University, Mumbai 400019, India

Abstract—Polyurethane dispersions (PUDs) synthesized using varying amounts of butyl acrylate (BA), from 0 i.e. with only NMP as the solvent to 50 wt%, were used as adhesives for laminates, made from PET and LDPE films. These adhesives were characterized by particle size analysis and contact angle measurements, while the graft efficiency of BA on PU was evaluated by FTIR and solvent extraction techniques. The effect of the ratio of BA to PUD on the contact angle and graft efficiency of acrylic monomer onto PU backbone was studied from the peel strength for plastic laminates. It was observed that with increasing BA content in the PUD, both particle size and graft efficiency decreased. These adhesives exhibited very good peel strength for the PET-LDPE laminates, fair for PET-PET and poor for LDPE-LDPE laminates. A strong relation was found between the graft efficiency of the BA and the adhesive characteristics of the PUD. At 40 wt% of BA in PUD the graft efficiency was the highest, while a high contact angle and low peel strength were observed for all plastics laminates. Therefore, lower graft efficiency is better in comparison to high grafting of BA in hybrid PUDs as an adhesive for plastic laminates.

Keywords: PUD; butyl acrylate; plastics laminates; PET; LDPE; adhesives; particle size analysis; contact angle; graft efficiency; FTIR; peel strength.

1. INTRODUCTION

During the last few years, waterborne polyurethanes (PUs) have begun to show intense interest in the application areas previously served only by solventborne resins due to two important reasons: one is the environmental concern, regarding VOC emission into the atmosphere, resulting into ozone depletion and ultimately an unbalance of Earth's ecosphere; and the second one is the economics, as an organic solvent is not an integral part of the adhesive which has a high weight proportion in the solventborne adhesive formulations, and this makes solventborne adhesives expensive. PUDs are free from the above-mentioned limitations; moreover, their

---

*Current address: Chemical & Materials Engineering Department, University of Alberta, Edmonton, Alberta, Canada.

[†]To whom correspondence should be addressed. E-mail: Jagtap@udct.org

performances have been improved up to the point that they are comparable to or better than the conventional solvent-based PUs for many applications [1].

One of the variants of solventborne resins is the low VOC PU possessing only a small amount of co-solvent introduced as a processing aid with an additional benefit of coalescing the particles in the dispersion into a smooth, continuous elastomeric film. N-methyl-2-pyrrolidone, (NMP) has been the co-solvent of choice. NMP is an expensive solvent and has a strong tendency to oxidize. The residual NMP in the dried film tends to cause yellowing, particularly at high temperatures and is capable of dissolving or swelling a wide spectrum of protective gears in an industrial environment. Moreover, worldwide legislation now stipulates re-labeling of all products containing NMP [2]. California Proposition 65 and European legislation require special labeling for products containing NMP; In Europe, products containing more than 5% NMP, have to be labeled as toxic and irritant, whereas worldwide trend in the paint industry is to eliminate NMP from all coating formulations.

Polyurethane dispersions (PUDs) can be customized to offer high performance in terms of their combination of toughness, abrasion resistance, flexibility and chemical resistance [3]. But, at the same time, they have disadvantages of high cost, low pH stability and limited outdoor durability; whereas, on the other hand, an acrylic emulsion has excellent weather resistance, good pigmentability and lower cost. So, in order to obtain cost/performance balance acrylic emulsions have been incorporated into the PU dispersions [4, 5]. Due to limited compatibility between these two systems, a physical blend of an acrylic emulsion and a PUD results in films with distinct phases of the two polymers [6].

Hirose *et al.* [7] used core-shell polymerization technique to obtain waterborne dispersions where acrylic monomers were polymerized after their diffusion into the bulk phase of PU polymer, to yield an interpenetrating polymer network (IPN) which, in turn, prevented phase separation. Kukunja *et al.* [4] have compared the performance of Acrylic-PU hybrid emulsions with physical blends of emulsions of an acrylic-PU system and found that hybrid emulsions provided improved performance for coatings application. Similar work was carried out by Brown *et al.* [8], in which the acrylic emulsion consisted of copolymer of methyl methacrylate/butyl acrylate/ acrylic acid; whereas PUDs were made of polyester polyol and aliphatic diisocyanate, and these dispersions were anionically stabilized, and consisted of 26% NMP as cosolvent. Hourston *et al.* [9] have reported the mechanical properties and morphology of PU/Polystyrene interpenetrating polymer networks for use in the field of coatings. Sebenik *et al.* [10] have synthesized Acrylic-PU hybrid emulsions by polymerization of acrylic monomers mixtures (BA, MMA, AA) in presence of PUDs (anionically stabilized, aliphatic polyester polyol based in NMP medium). The mechanical properties were reported for various emulsions prepared by varying the weight ratio of acrylic and PU components [10].

Most of the works reported on Acrylic-PU hybrid emulsions are in the form of patents [11–19] which have potential applications in the field of coatings and only a few publications [20–22] are available showing a systematic study. In all the above studies, the commercially available PUDs were used which already contained 15–26 weight% NMP as a cosolvent; acrylic components were mixtures of methyl methacrylate/butyl acrylate/acrylic acid monomers and comparison of properties of hybrid emulsions with physical blends was reported. Moreover, all the dispersions reported have been shown to have the potential for use as coatings. However, no work has been reported to show the graft efficiency of acrylic component on the PU backbone and its use as adhesives.

In the present study, anionically stabilized, aliphatic polyester based PUDs (completely NMP free) with varying fractions of n-BA were synthesized to obtain Acrylic-PU hybrid emulsions. n-BA included in the formulations acts as a diluent, lowering the glass transition temperature and providing tack for pressure sensitive adhesives.

The objective of this study was to investigate the effect of the amount of acrylic monomer n-BA on the wettability, graft efficiency and the final adhesive properties. Finally, the adhesive properties of Acrylic-PU hybrid emulsions were also compared with those of conventional NMP-containing PU dispersions.

2. EXPERIMENTAL

2.1. Materials

Adipic acid [DuPont], 1,2-propane diol, N-methyl-2-pyrrolidone, triethyl amine (TEA), ethylene diamine (EDA) as a chain extender [all materials from Merck], tetra-n-butyl titanate (TNBT) as a polyester catalyst [Super Urecoat Ind., India], isophorone diisocyanate (IPDI) [Degussa], dibutyl tin dilaurate (DBTL) [Dura Chemicals, India], n-butyl acrylate (n-BA) [National Chemicals, India], azobisisobutyronitrile (AIBN) as initiator [DuPont], were used as received and dimethylol propionic acid (DMPA) [Perstorp Co., India] was vacuum dried at 80–85°C before use.

2.2. Synthesis of polyester polyol [poly(propylene glycol) adipate]

The recipe for the preparation of the polyester polyol is given in Table 1. The reaction was carried out under an inert nitrogen atmosphere in a glass reactor, equipped with a fractionation column, a stirrer, and a thermometer.

All the reagents i.e. adipic acid, 1,2-propanediol, TNBT were charged into the reactor and slowly heated to 250°C under a nitrogen blanket. The progress of the reaction was monitored by measuring the water of reaction and the acid value. The reaction was stopped by cooling the reaction mixture when the acid value was below 2 mg KOH/g. At this stage, the excess glycol was removed by vacuum distillation.

242   *S. K. Patel* et al.

Table 1.
Recipe for the synthesis of polyester polyol

Component	No. of moles
Adipic Acid	14.00
1,2-Propanediol	16.58
TNBT (Catalyst)	4.20 (g)

Table 2.
Formulation for the synthesis of urethane prepolymer

Component	No. of moles	
	Set – 0	Sets – I to IV
IPDI	0.467	1.297
DMPA	0.139	0.386
Polyester polyol [Poly(propylene glycol)adipate]	0.157	0.436
n-Butyl acrylate	–	20% of total solids
NMP	15% of total solids	–

The final OH-terminated polyester polyol (poly(propylene glycol) adipate) had the following characteristics: Acid value: 1.20 mg KOH/g; OH value: 77 mg KOH/g; Viscosity: 4800 cp.

2.3. Synthesis of urethane prepolymer

The recipe for the synthesis of urethane prepolymer is shown in Table 2. The reactor was charged with IPDI, DMPA and urethane catalyst DBTL and the temperature was maintained at 80°C with continuous stirring with a loop stirrer until a homogeneous mixture was obtained.

This was followed by the addition of polyester polyol for further reaction at 80°C. In the hybrid emulsions, varying amounts of n-BA were added while in the case of conventional PUD, NMP was added. The reaction was monitored by change in the NCO value determined using dibutylamine back titration method [18]. The reaction was stopped when the desired NCO value of 3.7% was obtained.

2.4. Emulsification and chain extension

In hybrid emulsions, n-BA monomer was added from 20 to 50 wt% based on the solids as shown in Table 3. These polymers were dispersed in water by adding dropwise with high-speed agitation by maintaining the pH into distilled water at 60°C which already contained the stoichiometric amount of TEA. The remaining free isocyanate was reacted with the chain extender EDA to polyurea.

Table 3.
Formulation of NMP-free PUDs with varying amounts of n-BA as reactive diluent

Component	Amounts (g)				
	Set – 0	Set – I	Set – II	Set – III	Set – IV
PU Prepolymer	362	280	280	280	280
Distilled water	673	520	567	617	654
TEA	15	10.5	10.5	10.5	10.5
EDA	0.2	0.2	0.2	0.2	0.2
Total n-BA present	0.0	46.8	71.2	95.3	117.1

Table 4.
Effect of variation in the amount of n-BA as a reactive diluent on the properties of hybrid PUDs

Sample	BA : PU ratio (%)	Total BA present (g)	Grafting efficiency (%)	Weight fraction of BA grafted (g)	Weight fraction of BA as IPN (%)	Viscosity @ 30°C (cP)
Set – 0 (NMP)	0	0.00	–	–		17
Set – I	20	46.80	68%	31.80	15	40
Set – II	30	71.16	64%	45.50	26	34
Set – III	40	95.28	58%	55.20	40	26
Set – IV	50	117.09	41%	48.00	69	36

2.5. Preparation of acrylic-PU hybrid emulsions

For the preparation of n-BA-PU hybrid emulsions, monomer soluble initiator AIBN (0.7 wt% of the monomer) was added to the emulsions and polymerized under inert atmosphere at 75°C. During the polymerization, some of the n-BA gets grafted onto the backbone of PU chains on available graft sites, while the remaining n-BA gets homopolymerized and remains in the dispersions in the form of IPN.

The PUD of Set – 0 was prepared in the same manner as shown in Tables 2 and 3, except that NMP was used as co-solvent instead of n-BA. The solid content of the final emulsions was maintained around 35 ± 1 wt%.

2.6. Acrylic-PU hybrid emulsion characterization

2.6.1. Viscosity
The viscosity of all the emulsions was determined using a Brookfield LV viscometer with spindle No. 2 at 50 rpm at 30°C. The viscosity values are given in Table 4.

2.6.2. Measurement of graft efficiency using the solvent extraction method
A weighed amount of polymer was inserted in an end-over-end tumbler mixture at room temperature containing 100 ml toluene, a solvent for the poly(butyl acrylate) PBA, in the hybrid PUD. The solvent extraction was performed at room temperature for 24 hr to check the graft efficiency, GE. The weight of dissolved polymer was

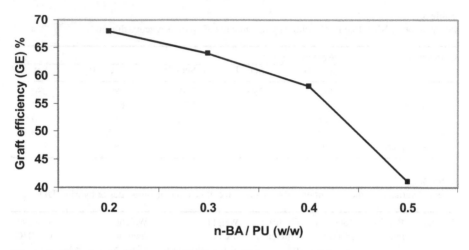

Figure 1. Plot of graft efficiency (%) of n-BA onto PU versus n-BA/PU ratio.

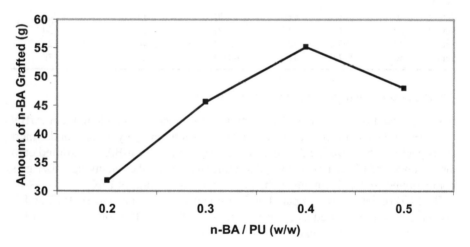

Figure 2. Plot of amount of n-BA grafted (g) versus n-BA/PU ratio.

determined from the solid content in the supernatant. Table 4 shows the data on graft efficiency. The plots of GE and the amount of BA grafted onto the PU backbone versus BA/PU ratio are shown in Figs. 1 and 2, respectively.

2.6.3. FTIR analysis
Infrared spectra of very thin and completely dried films were recorded using a Fourier Transform Infrared (FTIR) Spectrometer (SHIMADZU, model 8400-S). The film was characterized by the FTIR technique with 4 cm^{-1} resolution in the

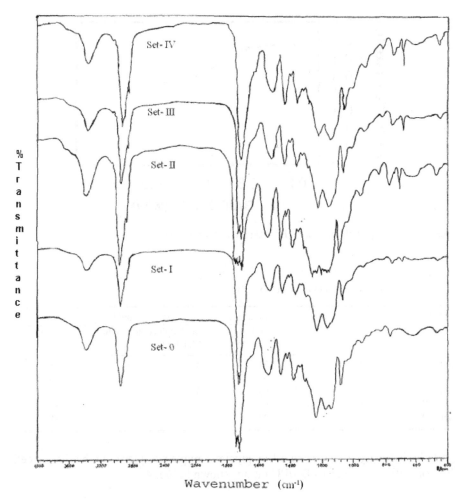

Figure 3. FTIR spectra of hybrid PUDs with different amounts of n-BA and PUD with NMP cosolvent. Set-0 – Hybrid PUD with no BA, but containing NMP cosolvent. Set-I – Hybrid PUD with 20 weight% BA. Set-II – Hybrid PUD with 30 weight% BA. Set-III – Hybrid PUD with 40 weight% BA. Set-IV – Hybrid PUD with 50 weight% BA.

mid-IR range from 4000 to 400 cm^{-1}. The spectra of the hybrid PUD and PUD with NMP are shown in Fig. 3.

2.6.4. Particle size analysis
The particle size was determined using COULTER LS 230 particle size analyzer on the weight percentage statistic. There was no special preparation of the sample for

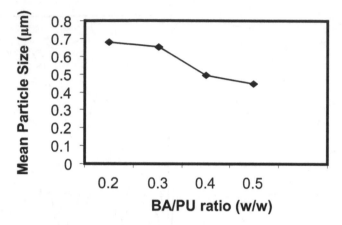

Figure 4. Plot of particle size versus BA/PU ratio in hybrid PUDs.

Table 5.
Contact angles of different liquids on various substrates

Substrate	Butyl acrylate (%)	Contact angle using different liquids (°)		
		Water	Low molecular weight PET liquid	Paraffin oil
Set – 0	0	77	32	33
Set – I	20	91	61	30
Set – II	30	93	62	29
Set – III	40	63	66	36
Set – IV	50	82	62	31

particle size analysis. By using a 1 ml pipette, sample was placed in the sample cell. Plot of particle size versus BA/PU ratio is shown in Fig. 4.

2.6.5. Contact angle measurements

Contact angles were measured using a KRUSS DSA10 contact angle meter. Drops of 0.06 to 0.07 ml liquids (water, low molecular weight liquid PET, and paraffin) were formed on the PUDs films with a micro-syringe. The equilibrium time for formation of liquid drop was 1 to 2 min. Contact angle was measured using a microscope having accuracy of ±2°, and values are reported in Table 5.

2.7. Laminate preparation

The adhesives developed in this study were intended to be used in flexible packaging industries. Poly(ethylene terephthalate) (PET) and low density polyethylene (LDPE) substrates (which are most commonly used substrates in such industries) were used to prepare laminates of PET to PET, PET to LDPE, LDPE to LDPE. The

Table 6.
Peel strengths of various laminates using different PUD hybrid adhesives

Sample	Butyl acrylate (%)	Peel strength (kN/m) PET/PET	PET/LDPE
Set – 0	0	98.1	196.2
Set – I	20	98.1	163.8
Set – II	30	111.8	144.2
Set – III	40	82.4	78.5
Set – IV	50	143.2	117.7

films of PET and LDPE were obtained from the Mittal Industries, India. All the films were wiped with isopropyl alcohol prior to lamination to remove foreign matter like dirt and dust particles and then subjected to a corona discharge (Sherman Treaters, Gujarat, India, G10 Corona Laboratory Bench Top unit with 280 mm wide U channel metallic electrode).

The adhesives were applied onto the corona treated side of the substrates. The adhesive was applied over the length, according to ASTM D 1876-72; with a bar coater which deposited a wet film of 24 μm thickness. After the adhesive had been applied to the first substrate, it was exposed to microwave radiation for one minute at 60–85°C in order to remove water from the adhesive film and then the second substrate was brought into contact with the adhesive film by applying a pressure of 260 kg/m^2. The laminates were then cut at the centre to produce 25 mm wide strips to avoid any edge effects.

2.8. Bond strength measurements

Laminated strips were peeled at 180° at a crosshead pull rate of 30 cm/min using an Instron tensile Model 3340 testing instrument. The bond strength was measured in kN/m. The results of peel strength are reported in Table 6.

3. RESULTS AND DISCUSSION

3.1. Viscosity

The viscosities of the PUDs are shown in Table 4. The viscosity of NMP based PUD was the lowest, indicating NMP was functioning in a better manner, compared to BA, in reducing the viscosity. The lower viscosity would impart better flow and leveling ability to the adhesive in achieving a good film. Whereas, in hybrid PUDs, the grafted BA increases the molecular weight and percent solids, which makes the polymer viscous. Perhaps, one can say that the higher the PBA in the IPN, more would be the viscosity and hence poorer will be the film property.

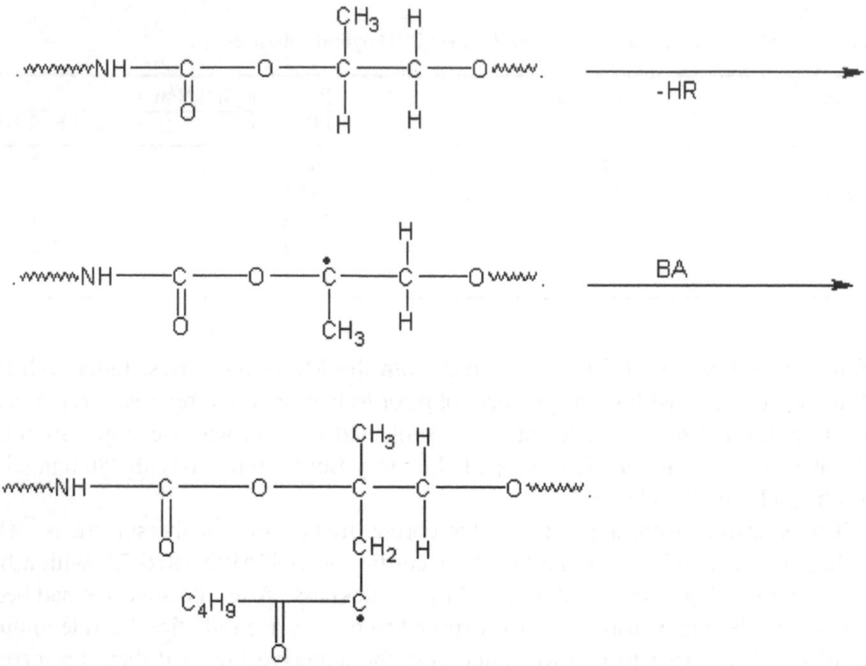

Figure 5. Schematic representation of grafting of butyl acrylate as a reactive diluent on polyester/urethanes (poly(propylene glycol)) by replacing labile H atom.

3.2. Effect of ratio of acrylate to urethane on graft efficiency (GE) in hybrid dispersions

n-BA does exhibit a high graft tendency, regardless of the type of the initiator used [19]. PU chains usually contain many α-hydrogen atoms from the poly(propylene glycol), present in polyester polyols. When n-BA is polymerized in presence of PUD, these α-H atoms can be attached by the free radicals formed either from the decomposition of the initiator or acrylic radicals, resulting in grafting of BA onto PU backbone. Thus the hybrid emulsion was composed of urethane/acrylic graft copolymer, ungrafted PU and acrylic polymer chains. The schematic diagram of graft reaction is shown in Fig. 5.

The GE and amount of n-BA grafted are shown in Table 4 and Figs. 1 and 2. It was observed that the graft efficiency decreased with the increase of n-BA concentration. As the concentration of n-BA increases, concurrently the concentration of PU decreases and as a result the number of graft sites decreases which, in turn, leads to a decrease in GE.

Figure 2 shows the actual amount of n-BA (in g) grafted onto the PU. With the addition of n-BA, a reactive diluent, the graft efficiency decreases continuously with the addition of n-BA in the emulsion. Although the grafted n-BA increases up to 40 wt%, the further addition decreases the grafted amount. In other words, the length of PBA grafted onto the PU backbone is the highest at 40 wt% n-BA. With the incremental addition of BA, as a reactive diluent, in PU the graft efficiency increased and was found to be highest for 40 wt% of n-BA, although it declines with further addition of n-BA. The higher graft efficiency results into higher number of n-BA monomers induced into the graft chains.

3.3. FTIR analysis

The IR Spectra of Sets – 0, I, II, III and IV are shown in Fig. 3. The infrared spectra of PUDs are mainly focused on two principal vibrational regions: the –NH stretching (3360 cm^{-1}) and C=O stretching (1700–1730 cm^{-1}). The IR spectra of the films showed no band at 2270 cm^{-1} which indicates the absence of free NCO and hence the completion of reaction.

The presence of bands at 1350 and 1180 cm^{-1} indicates the presence of alkyl compounds. Compared to Set – 0, in Sets – I, II, III and IV the area under the band at 2960 cm^{-1} (for C–H stretching) increased which also confirms the presence of long chain n-alkyl compounds. In acrylic-PU hybrid dispersions, the presence of band at 725 $\sim$ 720 cm^{-1} shows the presence of long chain n-alkyl compounds obtained after the polymerization of n-BA. As Set – 0 represents unmodified PUD in NMP, it lacks the band at 725 $\sim$ 720 cm^{-1}. Moreover, the area under the band at 725 cm^{-1} increases as the BA/PU ratio increases, which indicates the increase in chain length of PBA on the graft site and also confirms the increase in the extent of grafting of n-BA. Thus FTIR analysis reconfirms the results of graft efficiency obtained from the solvent extraction method.

3.4. Particle size analysis

The particle size analysis is an important criterion for the emulsions to be used in adhesive applications. The results reported in Fig. 4 show that as the % n-BA increases the particle size decreases. As the ratio of BA/PU increased the amount of PBA formed also increased. In short, in the PU-BA interpenetrating network the increased proportion of BA may trend to reduce the hydrodynamics volume of the PU dispersed particles.

3.5. Contact angle

The wetting property of the dried adhesive films was evaluated by measuring the contact angles of paraffin oil, low molecular weight PET liquid and water, exhibiting a range of polar character. The results are shown in Table 5. There was hardly any difference in the contact angle values for paraffin oil and they fall in range. Contact angles of polar, low molecular weight PET liquid increased with increase in the graft efficiency of butyl acrylate which reflects that the wetting and adhesion was poor;

whereas in case of water, the reverse was found i.e. the contact angle decreased with increase in graft efficiency. These results indicate that grafting made the adhesive more polar and caused difficulties in wetting and adhesion of relatively non-polar polymers like LDPE.

3.6. Peel strength

The results in Table 6 demonstrate that the PUD system with NMP as a co-solvent (Set – 0) showed good bond strength for both PET/PET and PET/LDPE laminates. For systems with acrylic modified PUDs, with n-BA/PU ratio of 0.2 and 0.3 (Set I and Set II), the performance of the adhesives is comparable to that of conventional solvent-based PU system. Moreover, as the n-BA/PU ratio increases, the peel strength keeps on decreasing up to a ratio of 0.4 (Set III) and then for the ratio of 0.5 (Set IV), the peel strength again increased.

In NMP based PUD, the urethane groups in the backbone have a tendency to form hydrogen bonds with the substrate, while the high boiling point NMP co-solvent helps in uniform film formation. As a result, Set – 0 exhibits good bond strength for the laminates.

In Sets I–IV, n-BA was used instead of NMP for the preparation of hybrid PUDs. In the hybrid PUDs, part of n-BA was homopolymerized and coexisted as IPN, whereas the remainder was grafted onto the PU backbone.

For a certain number of the pendent groups, in this case n-BA in hybrid PUDs, chain mobility increases, decreasing the glass transition temperature of the adhesive and also the tack which is important for pressure – sensitive adhesives ultimately increasing the bond strength of the laminates. Although, when the length increased beyond a certain number, the pendent groups start crystallizing, imparting stiffness to the polymer thus increasing the glass transition temperature, which reduces both the tack and bond strength of the laminates. Increased molecular weight enhances the cohesive strength, but it reduces the bond strength. In case of Set – I, where the amount of n-BA in IPN as well as grafted on the PU was the lowest of the remaining Sets, and this provides the overall best balance of adhesion and cohesion for the adhesive.

For non-polar liquids especially paraffin, all hybrid polymers, except Set III, exhibit low contact angles and the lowest is for Set – II i.e. 29° which is less than that for NMP-containing PUD. The contact angles for low molecular weight PET liquid are higher than NMP-containing PUD indicating relatively poor wettability of the adhesive on non-polar polymers compared to Set – 0 sample. Moreover, laminates after the peel test exhibited transfer of the adhesive on the relatively polar i.e. PET surface, indicating interfacial failure to the non-polar LDPE film. As the ratio of BA/PU increased the extent of homopolymer in the IPN continuously increased whereas the graft efficiency was highest for 40 wt% and further addition i.e. at 50 wt% the graft efficiency decreased. It was inferred that the lower the graft efficiency, the better was the bond strength. Therefore, for PET/PET laminates, n-BA/PU ratios of 0.2 and 0.5 offer the best results as an adhesive, whereas for PET/LDPE laminates n-BA/PU ratio of 0.2 offers the best results as an adhesive.

4. CONCLUSIONS

The NMP-free hybrid dispersions have been shown to provide adhesive properties comparable to their NMP-containing counterparts. The acrylic modified polyurethane dispersions showed good potential for use as laminating adhesives for flexible packaging films. With increasing the n-BA/PU ratio in the hybrid PUDs, the graft efficiency (GE) decreased with simultaneous increase of the BA graft chain onto PU, reducing the peel strength. It was observed that the lower the graft efficiency, the better was the bond strength. For PET/PET laminates, n-BA/PU ratios of 0.2 and 0.5 offer the best results as an adhesive; while for PET/LDPE laminates, nBA/PU ratio of 0.2 offers the best results as an adhesive. So, Acrylic-PU hybrid emulsion with n-BA/PU ratio of 0.2 offers the possibility to replace or partially replace the conventional PUD at a much lower cost, and meeting the industry demands.

REFERENCES

1. K. C. Frisch and D. Klempner (Eds.), *Advances in Urethane Science and Technology*, Vol. 10, p. 121 (1987).
2. I. Bechara, A. Ma and A. Puleo, *Paint & Coatings Industry*, **20**, 60 (2004).
3. M. Szycher, *Szycher's Handbook of Polyurethanes*, CRC Press, Boca Raton, FL (1999).
4. D. Kukanja, J. Golob, A. Zupancic-valant and M. Kranjc, *J. Appl. Polym. Sci.*, **78**, 67 (2000).
5. C. R. Hegedus and K. A. Kloiber, *J. Coatings Technol*, **68**(860), 39 (1996).
6. R. Satguru, J. McMahon, J. C. Padget and R. G. Coogan, *J. Coatings Technol.*, **66**(830), 47 (1994).
7. M. Hirose, F. Kadowaki and J. Zhou, *Prog. Organic Coatings*, **31**, 157 (1997).
8. R. A. Brown, R. G. Coogan, D. G. Fortier, M. S. Reeve and J. D. Rega, *Prog. Organic Coating*, **52**, 73 (2005).
9. D. J. Hourston, F. Schafer, N. J. Walter and M. H. Gradwell, *J. Appl. Polym Sci.*, **67**, 1973 (1998).
10. U. Sebenik, J. Golob and M. Kranjc, *Polym. Intl.*, **52**, 740 (2003).
11. T. E. Rolando, P. A. Voss and C. M. Ryan, U.S. Pat. 5,494,960 (1996).
12. E. Hansel, W. Meckel and T. Munzmay, U.S. Pat. 5,250,610 (1993).
13. H. L. Honig, G. Balle, W. Keberle and D. Dieterich, U.S. Pat. 3,705,164 (1972).
14. H. L. Honig, G. Balle, W. Kebrleand and D. Dietrich, U.S. Pat. 3,684,758 (1972).
15. S. Kaizerman and R. R. Aloia, U.S. Pat. 4,198,330 (1980).
16. B. R. Vijayendran, R. Derby and B. A. Gruber, U.S. Pat. 5,173,526 (1992).
17. P. Loewrigkeit and K. A. Van Dyk, U.S. Pat. 4,644,030 (1987).
18. H. C. Goos and, G. C. Overbeek, U.S. Pat. 5,137,961 (1992).
19. M. Guagliardo, U.S. Pat. 4,318,833 (1982).
20. A. Van den Elshout, *Surface Coatings International Part A*, **86**(A06), 229 (2003).
21. Y. He, E. S. Daniels, A. Klein and M. S. El-Aasser, *J. Appl. Polym Sci.*, **65**, 511 (1997).
22. E. C. Galgoci, C. R. Hegedus, F. H. Walker, D. J. Tempel, F. R. Pepe, K. A. Yoxheimer and A. S. Byce, *JCT CoatingsTech.*, **2**(13), 28 (2005).

23. D. J. David and H. B. Staley, *Analytical Chemistry of the Polyurethanes*, Vol. XVI, Part III, Robert E. Krieger Publishing Co. (1979).
24. K. C. Berger and Meyerhoff, in: *Polymer Handbook*, 3rd edition, J. Brandrup and E. H. Immergut (Eds.), John Wiley & Sons, New York (1989).

Contact Angle, Wettability and Adhesion, Vol. 5, pp. 253–267
Ed. K.L. Mittal
© VSP 2008

Cell adhesion to polystyrene substrates:
Relevance of interfacial free energy

ALAIN CARRÉ* and VALÉRIE LACARRIÈRE

Corning SAS, Corning European Technology Center, 7 bis, Avenue de Valvins, 77210 Avon, France

Abstract—Cell culture is one of the major tools used in cell and molecular biology. Most cells derived from solid tissues require an adhering surface to live and proliferate in vitro conditions. A good understanding of the relationships between the behavior of cells and the physicochemical properties of the substrates such as the surface free energy, the presence of functional groups and surface charges, is of prime importance for the optimization of adhesion, spreading and proliferation of cells.
Polystyrene and treated polystyrene surfaces were characterized by measuring the polar and non-polar components of the surface free energy of the substrates using wettability measurements. The adhesion of cells (Human embryonic lung fibroblasts, Chinese hamster ovary cells, Human umbilical vein endothelial cells) was quantitatively studied on different polystyrene substrates. In presence of bovine serum albumin, it is shown that the adhesion of cells increases with the polar component of the surface free energy. The situation is different in absence of serum, demonstrating the significant impact of the adsorption of biological compounds (proteins) on the adhesion of cells on solid substrates. This study shows that it is not directly the surface free energy of materials that controls cell adhesion but rather the interfacial free energy between the culture medium and the substrates. Therefore, we propose to consider the interfacial free energy between the culture medium and the substrates as a key parameter in determining cell adhesion and behavior on solid surfaces.

Keywords: Cell adhesion; surface free energy; surface polarity; interfacial free energy; cell culture; protein adsorption; polystyrene.

1. INTRODUCTION

Cell culture is one of the major tools used in cell and molecular biology. Culture of animal cell lines has many different applications including toxicity testing, cancer research, virology, cell-based manufacturing (production of vaccines, proteins, insulin, hormones, tissues and organs), gene therapy, cell therapy, and drug screening.

The understanding of the relationship between the behavior of living cells and the physicochemical properties of the culture substrate, such as wettability, presence of functional groups, surface charges and surface topography, is of prime importance

---

*To whom correspondence should be addressed. Tel.: 33 1 64 69 73 71; Fax: 33 1 64 69 74 55;
e-mail: carrea@corning.com

for the optimization of adhesion, spreading and proliferation of cells. Surface modification treatments such as plasma or corona oxidation have been investigated and shown to positively influence the cell adhesion and the cell growth processes. However, detailed or quantitative studies that establish clear relationships between the physicochemical properties of the culture system and the response of cells in the culture have been lacking. Data on the kinetics of the initial attachment process when cells first come into contact with the surface are scarce.

Under physiological conditions, the adhesion of mammalian cells to a material is mediated mainly by the presence of a protein adlayer [1]. Of major importance is the presence in the adlayer of extracellular matrix (ECM) proteins able to interact specifically with integrin receptors of the cell surface. The properties of the conditioning adlayer depend on the substrate, particularly on wettability and electrical properties of its surface, and on the composition of the liquid phase. Most cell culture media are indeed multi-component systems containing a wide range of proteins which may be involved in competitive or sequential adsorption.

Anchorage-dependent cells are cultivated in nutritive media supplemented with serum, necessary to support cell growth by provision of hormonal factors, attachment and spreading factors, and carrier proteins. A more complete description of the composition of fetal bovine serum (FBS) is given in Appendix 1. In this work, we give prominence to the effect of protein adsorption on cell adhesion to polystyrene (PS) substrates of different surface free energies. It appears that one major factor controlling the adsorption of serum components and subsequently controlling cell adhesion is the interfacial free energy between the culture medium and the solid substrate.

2. EXPERIMENTAL

2.1. Polymer substrates and biological materials

Three commercial sterile polystyrene (PS) products were considered: untreated PS (PS) (35 mm culture dish, Nalge Nunc International), tissue culture treated PS (TCT PS) (Costar® 6-Well Clear TC-Treated Microplates, Corning Inc.) and Corning® CellBIND® 6-Well Clear Microplate (Corning® CellBIND® Surface, Corning Inc.). The Corning® CellBIND® Surface is created using a patented plasma treatment to produce a surface that incorporates significantly more (50 to 60%) oxygen than traditional tissue culture surface treatments.

Three cell lines were chosen: Chinese hamster ovarian (CHO) cells, human fetal lung fibroblasts (MRC5) and human umbilical vein endothelial (HUVE) cells. CHO cells were seeded at a concentration of 200,000 cells per well in 2 ml Ham F-12 culture medium (Gibco® cell culture products, Invitrogen Corporation) containing 20 μl antibiotic (penicillin-streptomycin (10,000 units of penicillin (base) and 10,000 μg of streptomycin (base)/ml), Gibco® cell culture products, Invitrogen Corporation), with 10% or without FBS (Gibco® cell culture products, Invitrogen

Corporation) in the medium. MRC5 cells were seeded at a concentration of 50,000 cells per well in 2 ml of IMDM medium (Iscove's Modified Dubelcco's Medium, Gibco® cell culture products, Invitrogen Corporation) with 20 μl antibiotic, with 10% or without FBS in the medium. HUVE cells were seeded at a concentration of 90,000 cells per well in 2 ml of Ham F-12 medium containing 20 μl antibiotic, 0.1 mg/ml heparin (H3393, Sigma Aldrich), 0.05 mg/ml endothelial cell growth factor (ECGF, E2759, Sigma Aldrich) with 10% or without FBS in the medium. Adhesion experiments were conducted in presence of FBS, heparin and ECGF corresponding to the normal culture conditions or in absence of these components. After seeding, the cells were grown at 37°C in an incubator with 5% CO_2. The physiological pH of cell medium was 7.4.

The number of adhering cells was measured at the end of the adhesion step after 1.5 hr of incubation for the CHO cells and after 3 hr of incubation for the MRC5 and HUVE cells.

The methods of cell counting included manual counting with the Malassez cell, colorimetry (staining of cells with crystal violet) or flow cytometry (Z1Coulter Counter, Beckman).

2.2. *Contact angle measurements*

The surface free energy of polystyrene substrates, γ_S, was determined from contact angles, θ, of liquid probes of known surface tension, γ_L. By using the Fowkes theory [2], the surface free energy is split in a dispersion (non-polar) component, γ^D, and a polar component, γ^P. The geometric mean was used to calculate the work of adhesion, W. In the extended theory proposed by Owens and Wendt [3], the geometric means of non-polar and polar components are added so that the expression for the work of adhesion, W becomes:

$$W = 2 \left(\gamma_L^D \gamma_S^D \right)^{1/2} + 2 \left(\gamma_L^P \gamma_S^P \right)^{1/2} = \gamma_L (1 + \cos \theta) \tag{1}$$

The dispersion (non-polar) and polar components of the surface free energy of the polymer substrates, γ_S^D and γ_S^P, were determined from the contact angle of a non-polar liquid, tricresylphosphate (TCP, Sigma Aldrich), and from the contact angle of water. Water was purified by ionic exchange followed by organic removal leading to a resistivity of 18 MΩ.cm (Elgastat, UHP). With TCP (L), $\gamma_L = 40.9$ mJ/m^2 and $\gamma_L^D \approx 39.2$ mJ/m^2, so that the polar contribution in Equation (1) is neglected, and γ_S^D is deduced. Then the polar component, γ_S^P can be calculated from the water (W) contact angle knowing that, $\gamma_W = 72.8$ mJ/m^2, $\gamma_W^D = 21.8$ mJ/m^2 and $\gamma_W^P = 51$ mJ/m^2.

Alternatively, the polar contribution was deduced from the contact angle of water under octane (O) (reagent grade, Sigma Aldrich). For this particular system, the polar interaction I_{SW}^P is given [4, 5] as:

$$I_{SW}^P = 2 \left(\gamma_W^P \gamma_S^P \right)^{1/2} = \gamma_W - \gamma_O + \gamma_{WO} \cos \theta \tag{2}$$

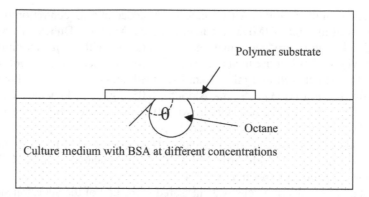

Figure 1. Method to measure the contact angle of octane in presence of an aqueous phase.

where γ_O (21.3 mJ/m^2) and γ_{WO} (51 mJ/m^2) are the octane surface tension, and the water/octane interface tension.

Contact angle measurements were made using a Ramé-Hart contact angle goniometer at a controlled temperature of 22 ± 1°C. The average contact angle was obtained from measurements on 5 liquid drops, measuring the contact angles on both sides of the drop. The liquid drops had a volume of 2 μl.

The adsorption of bovine serum albumin (BSA, a large globular protein of about 66 000 g/mol), the major protein of FBS, on PS, TCT PS and on the Corning® CellBIND® Surface was studied by measuring the contact angle, θ, of octane in presence of the culture medium containing different concentrations of BSA (BSA > 96%, ref A7888, Sigma Aldrich). As the plastic substrates have a density slightly lower than the culture medium, they were floating on the aqueous medium. As octane is less dense than water, the drop of octane was placed under the solid substrate as shown Fig. 1. The octane droplet (volume 2 μl) was formed with a microsyringe having a curved needle.

The concentration of BSA was varied from 10^{-5} g/l to 10 g/l. For each concentration, six contact angle measurements of octane were made.

2.3. Liquid/liquid interfacial free energy measurements

The interfacial tension between the culture medium and octane was measured with a Dynamic Contact Angle Analyzer (DCA, Cahn Instruments). Different amounts of BSA were added to the medium to study the adsorption of BSA at the medium/octane interface. The measurements were made at 22 ± 1°C.

2.4. Staining of proteins

The presence of proteins on polystyrene substrates can be revealed by staining of proteins with a colloidal gold dispersion (Colloidal Gold Total Protein Stain, Bio-

Rad Laboratories, Hercules, CA). The samples exposed to proteins were dipped in the gold staining dispersion for 18 hours, after which they were rinsed with pure water and blow-dried under nitrogen. The optical density of the gold coating was measured at 550 nm. The optical density increases with the amount of proteins retained on the solid substrates.

3. RESULTS

3.1. Surface free energy measurements

The contact angles of TCP, water and water in octane are presented in Table 1. The dispersion and polar components of the surface free energies deduced from Equations (1) and (2) are given in Table 2. The dispersion components of PS, TCT PS and of the Corning® CellBIND® Surface are comparable, in the range of 41–42 mJ/m^2. The polar components are very different increasing from PS to TCT PS and to the Corning® CellBIND® Surface. The polar components were deduced either from Equation (1) and the contact angles of TCP and water, or directly from Equation (2) and the contact angle of water in octane. The two methods of determination of γ_S^P are independent but lead to very close values. Surface polarity increases in the order: PS ($\gamma_S^P \approx 2$ mJ/m^2) < TCT PS ($\gamma_S^P \approx 12$ mJ/m^2) < Corning® CellBIND® Surface ($\gamma_S^P \approx 34$ mJ/m^2).

The total surface free energy of PS ($\gamma_S = \gamma_S^D + \gamma_S^P \approx 43$ mJ/m^2) is in good agreement with the value of 42 mJ/m^2 obtained by Wu [6] who used Equation (1) and the contact angles of water and methylene iodide on polystyrene. The surface treatments of TCT PS and of the Corning® CellBIND® Surface undoubtedly increase very significantly the surface polarity of the polymers, leaving the dispersion component of PS almost unaffected.

Table 1.

Contact angles (°) measured on the different polystyrene substrates

Materials	θ of TCP	θ of water	θ of water in octane
PS	17.0 ± 1.1	85.1 ± 10.9	127.5 ± 0.3
TCT PS	12.6 ± 1.0	60.0 ± 0.5	92.9 ± 6.9
Corning® CellBIND® Surface	16.6 ± 1.0	31.0 ± 1.7	46.0 ± 2

Table 2.

Surface free energy components (mJ/m^2) of the different polystyrene substrates

Materials	γ_S^D (from θ of TCP)	γ_S^P (from θ of TCP and water)	γ_S^P (from θ of water in octane)
PS	41	1.9	2.0
TCT PS	42	11.7	11.6
Corning® CellBIND® Surface	41	30.5	36.6

3.2. Cell adhesion

Following the protocol developed by Vogler and coworkers [7, 8], the steps of cell adhesion and proliferation were clearly identified for each cell line. The adhesion step is typically of a few hours and precedes the exponential proliferation step. Cell proliferation stops at confluence, i.e., when the entire available surface of the substrate is covered by the cells.

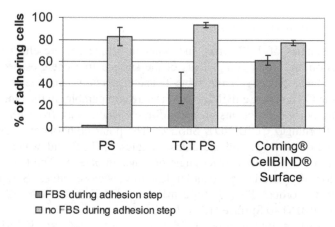

Figure 2. Percentage of MRC5 adhering cells on PS, TCT PS and the Corning® CellBIND® Surface with and without FBS in the culture medium, after 3 hours of incubation.

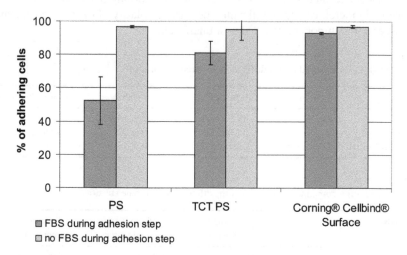

Figure 3. Percentage of CHO adhering cells on PS, TCT PS and the Corning® CellBIND® Surface with and without FBS in the culture medium, after 1.5 hours of incubation.

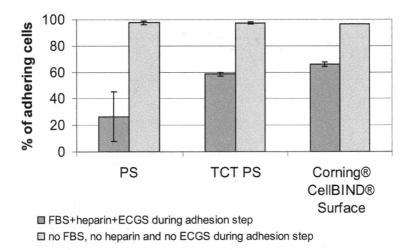

Figure 4. Percentage of HUVE adhering cells on PS, TCT PS and the Corning® CellBIND® Surface with and without FBS in the culture medium, after 3 hours of incubation.

Figures 2, 3 and 4 show the percentage of adhering cells on PS, TCT PS and on the Corning® CellBIND® Surface in presence or absence of FBS at the end of the adhesion step. Cell adhesion appears to be greatly dependent on the presence or absence of serum (FBS) in the culture medium. In presence of bovine serum albumin, the number of adhering cells increases with the polar component of the surface free energy for all three cell lines. The percentage of adhering cells is notably increased in absence of FBS, especially on the PS substrate.

It was already reported that the presence of serum in culture medium may inhibit or delay cell adhesion compared to serum-free medium [1, 9–13]. In many cases, it seems likely that cell adhesion in the presence of serum is governed by competition between adsorption of ECM proteins secreted by the cells or already present in the serum, mainly fibronectin, vitronectin, and collagen, and adsorption of the two most abundant proteins in serum, albumin and globulin.

Albumin and globulin have the highest concentrations in serum [11]. As bovine serum albumin (BSA) makes up approximately 50% of the total protein in fetal bovine serum, it is usually thought that BSA can be easily adsorbed on hydrophobic substrates such as PS, leading to a "passivation" of the surface. To evaluate this assumption, the adhesion of CHO cells on PS was quantified in the culture medium supplemented only by BSA.

The adhesion of CHO cells on PS was characterized by counting the number of adhering cells on PS as a function of the BSA concentration in the medium after 1.5 hours of incubation. The results are reported in Fig. 5. The pictures in Fig. 6 show the density of adhering CHO cells on PS as a function of the BSA concentration (0, 0.01, 0.1, 0.3 g/l). It appears that the surface density and the morphology of

A. Carré and V. Lacarrière

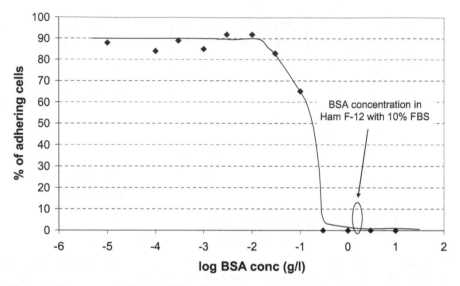

Figure 5. Percentage of adhering CHO cells on PS as a function of the BSA concentration in the culture medium.

cells are different depending on the BSA concentration. The first consequence of adding BSA to the medium below 0.01 g/l is that cells spread less on the substrate without reducing the number of adhering cells as shown in Figs. 5 and 6. Above a concentration of 0.01 g/l, the number of adhering cells is reduced (Fig. 5). At a concentration of 1.5 g/l of BSA in the culture medium supplemented with 10% of FBS (FBS contains 15 g/l of BSA), no cells adhere to PS. However, with a medium supplemented with 10% of serum, the percentage of adhering CHO cells is still 45% (Fig. 3). Thus, in the current culture medium (with 10% of FBS) it seems that the negative impact of BSA is somehow balanced by other components of FBS, promoting cell adhesion. One of those could be fibronectin which is an important protein for cell adhesion to substrates in spite of its quite low concentration in the serum (0.2 g/l (11)).

3.3. Adsorption of BSA on substrates

The adsorption of BSA on PS was studied by measuring the contact angle of an octane drop in presence of the culture medium containing different concentrations of BSA (Fig. 1). One relevant thermodynamic function describing the equilibrium of a drop on a solid is the cosine of the contact angle. Figure 7 represents the variation of the cosine of the contact angle of octane on PS, TCT PS and the Corning® CellBIND® Surface in Ham F-12 medium as a function of the BSA concentration in the medium. The results evidence strong adsorption on PS from a low concentration of 3.10^{-4} g/l. The decrease of the cosine of the octane contact

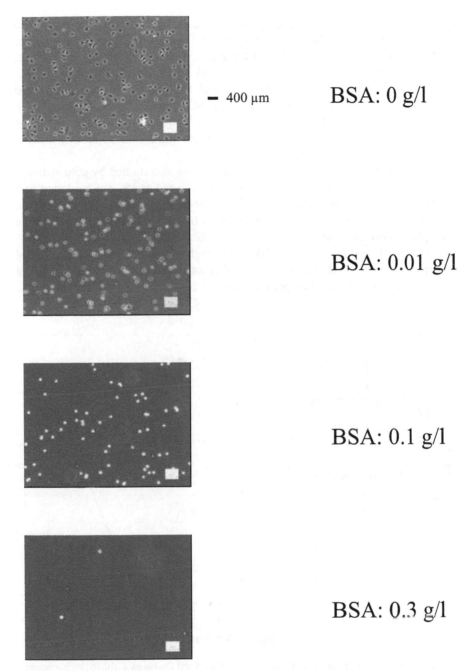

Figure 6. Density of adhering CHO cells on PS as a function of the BSA concentration (incubation period: 1.5 hrs).

angle may be attributed to the adsorption of species modifying the interfacial free energies between the medium and the substrate, between the medium and octane, and the substrate and octane. On TCT PS and the Corning® CellBIND® Surface, BSA adsorption is also visible but from a BSA concentration of 3.10^{-2} g/l, i.e. 100 times higher than with PS. The interfacial free energy between octane and Ham F-12 decreases when the BSA concentration increases (Table of Appendix 2). Therefore, the increase of the contact angle (Fig. 7) indicates a stronger reduction of the interfacial free energy between the medium and the substrate due to BSA adsorption, especially with PS.

The adsorption of BSA on the three substrates was also studied by gold staining. In Fig. 8, the optical density of the gold stain measured at 550 nm is plotted as a function of the log of the BSA concentration in the medium (Ham F-12). Adsorption of BSA on PS is detected from a concentration of 10^{-4} g/l of BSA. On the more hydrophilic substrate (the Corning® CellBIND® Surface) the adsorption of BSA occurs from 10^{-2} g/l, i.e. again 100 times higher than with PS. At the actual concentration of BSA in the medium supplemented with 10% of FBS (1.5 g/l of BSA), the amount of adsorbed protein seems to be about twice on PS than on the

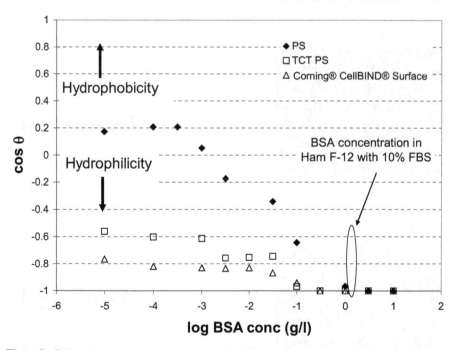

Figure 7. Cosine of the octane contact angle on PS, TCT PS and Corning® CellBIND® Surface as a function of the concentration of BSA in the Ham F-12 medium.

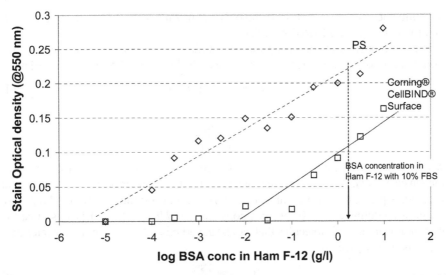

Figure 8. Adsorption of BSA on polystyrene substrates. The protein was stained with colloidal gold dispersion and the optical density was measured at 550 nm.

Table 3.
Interfacial free energy γ_{SW} (mJ/m^2) between water and the different polystyrene substrates

Materials	γ_{SW}
PS	36.0
TCT PS	17.0
Corning® CellBIND® Surface	5.0

Corning® CellBIND® Surface. TCT PS data (not shown for clarity) lie between the data of PS and the Corning® CellBIND® Surface.

From these simple experiments, it can be concluded that some FBS components such as BSA are adsorbed on cell culture substrates. This adsorption of these compounds may be responsible for a reduction of the amount of adhering cells, especially on the more hydrophobic substrate (PS).

The adsorption of molecules at interfaces is a thermodynamically driven process. The adsorption leads to a decrease of the free energy of the system. The interfacial free energy between water and the different polystyrene substrates can be deduced from the surface free energies of water and of the solids according to [3].

$$\gamma_{SW} = \gamma_S + \gamma_W - 2\sqrt{\gamma_S^D \gamma_W^D} - 2\sqrt{\gamma_S^P \gamma_W^P} \qquad (3)$$

The values of the interfacial free energies between water and the polystyrene substrates are given in Table 3. The interfacial free energies between water and the solid substrates are 36.0, 17.0 and only 5.0 mJ/m^2, respectively, for PS, TCT PS and the Corning® CellBIND® Surface.

The thermodynamics of adsorption of a surface active compound at a concentration c at the solid/water interface can be described by the Gibbs equation [14] as:

$$d\gamma_{SW} = -RT\Gamma d\ln c \qquad (4)$$

or

$$\Gamma = \frac{-d\gamma_{SW}}{RTd\ln c} \qquad (5)$$

where R is 8.32 J/Mol.K, T is temperature (in K), Γ is the excess concentration of the surface active molecules at the interface, i.e. the number of moles of surfactant molecules per unit area present at the interface. For a dilute solution, taking pure water as a reference ($c = 0$), we can consider that $d\ln c = \frac{dc}{c} \approx 1$ [15], so that the maximum amount of molecules that can be adsorbed at the water/polymer interface, Γ_{max}, is simply given by:

$$\Gamma_{max} \approx \frac{\gamma_{SW}}{RT} \qquad (6)$$

Therefore, the maximum amount of molecules of a surface active compound that can be adsorbed at the substrate/water interface is proportional to the interfacial free energy, γ_{SW}. A direct proportionality between the interfacial free energy and the capacity for adsorption at an interface was already proposed by Lucassen-Reynders with usual surfactants [16]. This simple argument may explain why BSA is more adsorbed on PS ($\gamma_{SW} = 36.0$ mJ/m^2) than on TCT PS ($\gamma_{SW} = 17.0$ mJ/m^2) and than on the Corning® CellBIND® Surface ($\gamma_{SW} = 5.0$ mJ/m^2).

Surface properties of substrates may also have an important influence on the conformation of proteins and on the competition between the adsorption of different proteins as it was already reported [11, 17]. To complete the description of the protein adsorption on solid surfaces, the next problem to address is the competition between proteins inhibiting and promoting cell adhesion as a function of the surface properties of the different substrates, including the influence of surface charges. Concerning this last point, as BSA (inhibiting cell adhesion) and fibronectin (promoting cell adhesion) have isoelectric points of about 5 (i.e. they are negatively charged at pH = 7.4), and as PS surfaces are also usually negatively charged [17], an electrostatic repulsion between the proteins and the substrates can be expected. However the relatively high concentration of salt in a culture medium (about 0.12 mol/l) can inhibit electrostatic interactions as the screening or Debye length is less than 1 nm at this salt concentration [18].

4. CONCLUSION

Cell adhesion is a very complicated process. Adhesion of anchorage-dependent mammalian cells is traditionally viewed through at least four major steps that precede proliferation: protein adsorption, cell-substrate contact, cell-substrate attachment, and cell spreading. Protein adsorption is itself a complex process involving molecular-scale interactions with the substrate in presence of the aqueous culture medium.

It has been shown that the presence of FBS in a culture medium has a strong negative influence on the adhesion of cells, especially on hydrophobic PS. In absence of FBS during the adhesion step of cells, the number of adhering cells on PS, TCT PS and on the Corning® CellBIND® Surface is dramatically increased to reach 90% or more of adhering cells (nevertheless, serum is necessary for the proliferation of cells).

The determination of the surface free energy components of the culture substrates gives a clearer view of the influence of the surface properties on cell adhesion. However, this study shows that it is not directly the surface free energy of materials that controls cell adhesion but rather the interfacial free energy between the culture medium and the substrate. The interfacial free energy between the culture medium and the solid surface controls the adsorption of serum components that may inhibit cell adhesion. One of these components is likely to be BSA. The results indicate that BSA adsorption is related to the interfacial free energy between the medium and the substrate, and that BSA strongly inhibits cell adhesion. Hydrophilic substrates, such as TCT PS and especially the Corning® CellBIND® Surface, have a lower interfacial free energy with water than hydrophobic PS leading to a lower adsorption of protein inhibiting cell adhesion. Therefore, we propose to consider the interfacial free energy between the culture medium and the substrate as a key parameter in determining cell behavior on solid surfaces.

REFERENCES

1. J.-L. Dewez, V. Berger, Y.-J. Schneider and P. G. Rouxhet, *J. Colloid Interface Sci.* **191**, 1 (1997).
2. F. M. Fowkes, *Ind. Eng. Chem.* **56**(12), 40 (1964).
3. D. K. Owens and R. C. Wendt, *J. Appl. Polym. Sci.* **13**, 1741 (1969).
4. A. Carré, V. Lacarrière and W. Birch, *J. Colloid Interface Sci.* **260**, 49 (2003).
5. A. Carré and V. Lacarrière, in *Contact Angle, Wettability and Adhesion*, K. L. Mittal (Ed.), Vol. 4, pp. 267-280, VSP/Brill, Leiden (2006).
6. S. Wu, *J. Polym. Sci.* **C34**, 19 (1971).
7. E. A. Vogler and R. W. Bussian, *J. Biomed. Mater. Res.* **21**, 1197 (1987).
8. J. Y. Lim, X. Liu, E. A. Vogler and H. J. Donahue, *J. Biomed. Mater. Res.* **68A**, 504 (2004).
9. Y. Tamada and Y. Ikada, *J. Colloid Interface Sci.* **155**, 334 (1993).
10. J. A. Witkowki and W. D. Brighton, *Expl. Cell Res.* **70**, 41 (1972).
11. F. Grinnel and M. K. Feld, *J. Biol. Chem.* **257**, 4888 (1982).
12. A. S. G. Curtis and J. V. Forrester, *J. Cell Sci.* **71**, 17 (1984).
13. P. Knox, *J. Cell Sci.* **71**, 51 (1984).
14. M. J. Rosen, *Surfactants and Interfacial Phenomena*, 2nd edition, John Wiley & Sons, New York (1989).
15. V. Thoreau, L. Boulangé and J. C. Joud, *Colloids Surfaces A* **261**, 141 (2005).
16. E. H. Lucassen-Reynders, *J. Phys. Chem.* **67**, 969 (1963).

17. J.-L.Dewez, A. Doren, Y.-J. Schneider and P. G. Rouxhet, *Biomaterials* **20**, 547 (1999).
18. J. Israelachvili, *Intermolecular and Surface Forces*, second edition, p. 238, Academic Press (1992).
19. C. E. A. Jochems, PhD Thesis, Department of Laboratory Animal Science, Utrecht University, and Department of Animal Sciences, Wageningen Agricultural University (1997).

APPENDIX 1. COMPOSITION OF THE FETAL BOVINE SERUM

Fetal bovine serum (FBS) is described as being the "most universally applicable cell culture additive for the stimulation of cellular proliferation and biological production". It is produced from the blood of bovine fetuses. When the blood has been removed from the fetus, the blood is chilled and allowed to clot. Then, the raw serum is decanted, pooled and quickly frozen. Many low and high molecular weight substances are present in FBS, though the exact composition is not known yet. Many substances are demonstrated to be present but the exact effect on the cultured cells is still unclear. The dissolved substances present in fetal bovine serum can be summarized as follows [19]:

- Proteins: transport proteins (such as albumin), proteins necessary for cell adhesion to the matrix or attachment factors (such as fibronectin), enzymes, cell-type specific growth factors (for example fibroblast growth factor (FGF) and epidermal growth factor (EGF)), growth inhibition factors, hormones (insulin, one of the hormones that regulates carbohydrate metabolism, follicle stimulating hormone and growth hormone).

- Non-protein hormones: steroid hormones and derived substances, such as hydrocortisone, cortisol and testosterone.

- Lipids: factors required for cell growth and differentiation (cholesterol and phospholipids), factors required for cell proliferation (prostaglandins E and $F_{2\alpha}$).

- Minerals: K^+, Na^+, Cl^-, Ca^{2+} and phosphate ions are present, as well as several trace elements (Fe^{2+}, Cu^{2+}, and Zn^{2+}).

- Metabolites and nutrients: free fatty acids, glucose and ketone bodies are just a few examples of the many other nutrients and intermediary metabolites present in FBS.

- Others: FBS also contains substances functioning as vitamins, and substances functioning in pH buffering, protease inhibition and binding, and inactivation of toxic materials.

FBS contains also many other unknown substances necessary for growing most mammalian cells so that supplementing the culture medium with FBS is a necessity for culturing cells in vitro.

APPENDIX 2. ADSORPTION OF BSA AT THE OCTANE/HAM F-12 CULTURE MEDIUM INTERFACE

The adsorption of BSA at the octane/Ham F-12 culture medium interface was studied by measuring the interfacial tension between the two media as a function of the concentration of BSA added to the Ham F-12 medium. The DCA allows measuring directly the interfacial tension between the aqueous phase and octane. The data are reported in Table A1.

The interfacial free energy between the Ham F-12 medium and octane is slightly lower than with pure water (51 mJ/m^2). Presence of organic compounds (amino-acids, vitamins, glucose, etc...) in the culture medium is probably responsible for the slight decrease.

Table A1.

Interfacial tension between octane and Ham F-12 culture medium as a function of the BSA concentration, c, in the Ham F-12 culture medium ($T = 22°C$)

BSA concentration, c (g/l)	Interfacial tension between octane and the Ham F-12 culture medium, γ_{OW} (mN/m)
0	42.3
0.0001	38.8
0.0003	37.5
0.001	33.2
0.003	29.8
0.01	24.8
0.03	20.8
0.1	18.8
0.3	20.8
1	17.6
3	14.9
10	14.8

Part 3

Superhydrophobic Surfaces

Contact Angle, Wettability and Adhesion, Vol. 5, pp. 271–278
Ed. K.L. Mittal
© VSP 2008

Superhydrophobic aluminum surfaces obtained by chemical etching

D. K. SARKAR and M. FARZANEH*

Canada Research Chair on Atmospheric Icing Engineering of Power Networks (INGIVRE) and Industrial Chair on Atmospheric Icing of Power Network Equipment (CIGELE) Icing Research Building, Université du Québec à Chicoutimi (UQAC) 555, Boulevard de l'Université, Chicoutimi, Québec G7H 2B1, Canada

Abstract—Chemical etching has been used to create nanostructured patterns on aluminum surfaces. The micro/nanostructure obtained of the surfaces has been examined using scanning electron microscopy (SEM) and atomic force microscopy (AFM). The nanostructured patterns have been passivated using stearic acid and fluoroalkylsilane molecules to obtain low energy surfaces terminated with $-CH_3$ and $-CF_3$ groups, respectively. Optimized annealing temperatures have been obtained for both systems to achieve the highest contact angle and the lowest hysteresis. These surfaces are found to be highly superhydrophobic with a water contact angle higher than 170° with a very low hysteresis. However, $-CF_3$ terminated surfaces show lower hysteresis as compared to $-CH_3$ terminated surfaces.

Keywords: Chemical etching; nanostructured patterns; superhydrophobicity; contact angle; thermal desorption.

1. INTRODUCTION

Inspired by nature, researchers have created many superhydrophobic surfaces where water drops roll off removing surface contamination as observed on the lotus leaf and some creatures in nature [1–3]. In general, the contact angle of water on smooth, flat hydrophobic surfaces, such as Teflon, does not exceed 120° [4]. However, the addition of roughness to the surface can increase the contact angle without altering the surface chemistry. The resulting superhydrophobicity is attributed to the combined effect of surface morphology and surface chemistry. These superhydrophobic surfaces exhibit a very high contact angle of water and a very low contact angle hysteresis. As a result, water drops roll off from such surfaces even with a slight inclination, making them very attractive to various technological areas, for example, microfluidic devices, textile industries, and possibly anti-icing applications. Erbil *et al.* [5] described a simple and inexpensive method

---

*To whom correspondence should be addressed. Tel.: 1-418-545-5011 ext. 5044; Fax: 1-418-545-5032; e-mail: farzaneh@uqac.ca

for forming a superhydrophobic coating using polypropylene and an appropriate selection of solvent and temperature to control the surface roughness. Recently, lithography has been used to create ordered structures on surfaces to demonstrate superhydrophobicity [6]. Onda *et al.* [7] achieved a contact angle of 174° on a surface using the fractal structure of alkylketene. Other methods include the sol-gel process to generate a porous rough surface [8], anodic oxidation of aluminum surfaces [9] and plasma polymerization [10]. In many of these cases, a rough surface was created initially, which was then coated with a thin hydrophobic film with low surface energy. Organic acids and fluoroalkylsilane (FAS) are widely used for passivation to obtain superhydrophobicity [11, 12]. However, no results are available on correlating the thermal desorption of these organic molecules with superhydrohobicity at elevated temperatures.

In this study we report the formation of nanostructured patterns on aluminum surfaces using chemical etching and compare the superhydrophobic properties of these surfaces by passivating with stearic acid (SA) and fluoroalkylsilane (FAS) organic molecules. We also demonstrate a strong correlation between the superhydrophobic behavior and the thermal desorption of these organic molecules bonded to the nanostructured surface at elevated temperatures.

2. EXPERIMENTAL

Aluminum surfaces were etched with 18.5 w% dilute hydrochloric acid (HCl) solution for a maximum of 4 minutes. All the etched samples were ultrasonically cleaned with deionized water to remove any residual dust particles from the pores of the nanosurface. The cleaned samples were dried at 70°C for more than 10 hours. Then, some of them were immersed in a stearic acid (SA) solution and others in an FAS solution for 30 minutes for passivation. The SA solution was prepared by dissolving 2×10^{-3} molar solid stearic acid (SA) in acetone. Similarly, the FAS solution was prepared by dilution in ethanol in the volume ratio of FAS:Ethanol=3.6:96.4. The chemical name of FAS is triethoxy-1H,1H,2H,2H-perfluorodecylsilane ($C_{16}H_{19}F_{17}O_3Si$) (product number 658758, Sigma Aldrich). A few silicon samples were passivated in a similar manner for Fourier Transform Infrared (FTIR) Spectroscopy investigation. The passivated aluminum and silicon samples were dried at 70°C for more than 2 hours. The SA and FAS passivated aluminum surfaces were annealed at different temperatures for 30 minutes to study the thermal stability of the hydrophobic coating. The surface morphology of the samples was investigated using field emission scanning electron microscopy (FEGSEM, Leo 1525) and atomic force microscopy (AFM) (Digital Nanoscope IIIa by Digital Instruments). The FTIR investigations of the passivated samples were carried out using a PerkinElmer Spectrum One FTIR spectrometer in the wavenumber range of 400–4000 cm^{-1}. Equilibrium contact angle (CA) measurements were made using a Krüss DSA100 goniometer.

3. RESULTS AND DISCUSSION

Figure 1 shows the FESEM micrograph of an aluminum surface etched with 18.5 w% HCl for 4 minutes. Cloudy fiber-like structure is observed due to the formation of etch pits during etching. The etch pits formation usually takes place at the dislocation sites of the aluminum surface and the number of etch pits is proportional to the concentration of the acid [13]. AFM has been used to investigate the details of the microstructural evolution due to the formation of etch pits. Figure 2(a) shows the tapping mode AFM image of the as-received aluminum surfaces at

Figure 1. FESEM micrograph of Al surface etched with 18.5 w% HCl acid.

Figure 2. AFM images of (a) unetched Al surface and etched with 18.5 w% HCl for (b) 2 minutes and (c) 4 minutes. The scan size of these images is 2 μm ×2 μm.

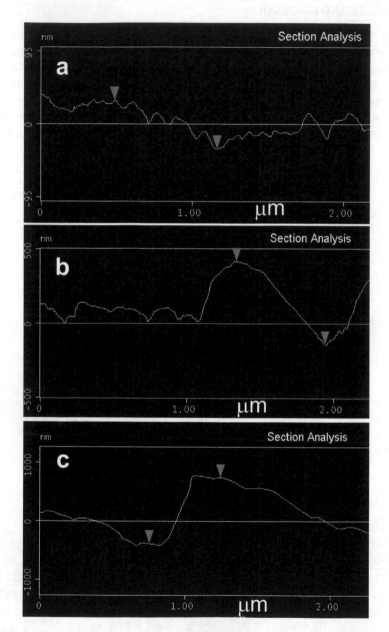

Figure 3. Section analysis of the AFM images of (a) unetched Al surface and etched with 18.5 w% HCl for (b) 2 minutes and (c) 4 minutes.

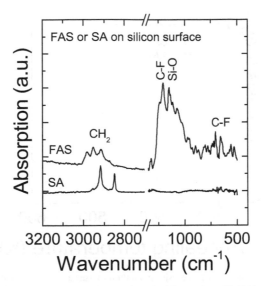

Figure 4. FTIR spectra of silicon surfaces passivated with stearic acid (SA) and fluoroalkylsilane (FAS).

a scan size of 2 μm $\times$2 μm with a root mean square (rms) roughness of $\sim$ 30 nm. The size of the individual grains is found to be $\sim$ 50 nm, which is evident from the section analysis shown in Fig. 3(a). The rms roughness of the surface produced using 18.5 w% acid etching for 2 minutes is 165 nm as shown in Fig. 2(b). Figure 3(b) shows the section analysis of Fig. 2(b) where the average grain size is found to be $\sim$ 300 nm which is further decorated with small grains of $\sim$ 100 nm. Increasing the etching time to 4 minutes increased the rms roughness to 366 nm with an average grain size of $\sim$ 700 nm, as shown in Fig. 2(c). The section analysis shows that each grain is further decorated with small grains of $\sim$ 100 nm (Fig. 3(c)). It is evident from Fig. 2(b) and Fig. 2(c) that the surface roughness achieved increases with the increase of the etching time and that the small grains are etched out leaving only the big grains on the surface.

These nanostructured aluminum surfaces were passivated with stearic acid (SA) and fluoroalkylsilane (FAS) molecules to produce low energy surfaces. Figure 4 shows the FTIR spectra of stearic acid (SA) and fluoroalkylsilane (FAS) passivated silicon surfaces. The peaks at 2848 and 2917 cm^{-1} are identified as the symmetric and asymmetric vibrations of $-CH_2-$ and $-CH_3$ groups of the stearic acid, respectively [14]. The most intense bands between 1050 and 1200 cm^{-1} are assigned to C–F stretching of $-CF_3$, $-CF_2-$ and Si–O–C stretching of FAS molecules [15].

It is found that the contact angle of water on both SA- and FAS-coated etched aluminum surfaces is higher than 170° and that the water drops roll off these

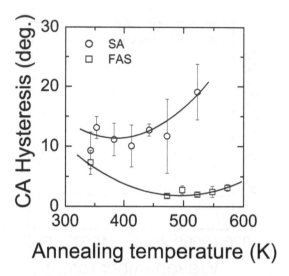

Figure 5. Contact angle (CA) hysteresis of aluminum surfaces etched with 18.5 w% HCl and passivated with stearic acid (SA) and fluoroalkylsilane (FAS) and annealed at different temperatures.

surfaces even with a slight tilt. Although contact angle remained nearly similar for both SA and FAS passivated surfaces, a drastic difference in contact angle hysteresis (CAH) was observed for the samples annealed at different temperatures. Figure 5 shows the contact angle hysteresis at different annealing temperatures for SA and FAS passivated surfaces created using 18.5 w% HCl acid for 4 minutes. The CAH values of the SA and FAS passivated surfaces were found to be 9.3° and 7.3°, respectively, for the samples dried at 343 K. SA and FAS passivations terminate the surface with $-CH_3$ and $-CF_3$ radicals, respectively. Usually, a $-CF_3$ terminated surface exhibits lower surface energy as compared to a $-CH_3$ terminated surface as the C-F bonds are stronger than the C-H ones due to higher electronegativity of F. Thus, the CAH is lower for a FAS terminated surface than an SA one. From Fig. 5, it is clear that the CAH of the SA passivated surface increases with the increase of annealing temperature, whereas for the FAS terminated surface the CAH decreases within the same range of annealing temperatures. At an annealing temperature of 473 K, CAH is reduced from 7.3° to 1.7° for the FAS terminated surface while CAH is increased to 14.3° from 9.3° for the SA terminated surface. This remarkable difference is due to the difference in thermal desorption rates between SA and FAS for aluminum surfaces at this temperature. FAS would produce covalent bonds, as silane is covalently bonded to oxidized surfaces [16, 17] and SA molecules are reported to be chemisorbed on oxidized or metallic surfaces [18, 19]; consequently, thermal desorption of SA molecules is easier than of FAS molecules. Therefore, a sufficient amount of SA molecules disappears from the surface, at 473 K, so as to effectively increase the CAH. On the other hand, the FAS molecules are presumably so well organized on the surface that no discernible thermal desorption occurs at this temperature, which effectively reduces the CAH. The CAH is increased to $\sim 20°$ for the SA passivated surface annealed at 523 K and water drops stick to the surface. It is reported that the thermal desorption of SA from nanostructured metallic surfaces takes place at 523 K [20]. Therefore, a large number of SA molecules would be lost by thermal desorption at this temperature so as to cause an increase in CAH. On further annealing at 573 K, the SA molecules may decompose completely leaving only traces of carbon on the aluminum surface, making it completely hydrophilic [21]. Previous studies have created superhydrophobic surfaces using FAS passivation; however, no systematic results have been reported on the behavior of contact angle and contact angle hysteresis with different annealing temperatures [11, 12]. Qian and Shen [12] and Suzuki *et al.* [11] annealed FAS passivated surfaces at 403 K and 523 K, respectively, for an hour to obtain superhydrophobicity. However, no clear justification has been made on the choice of annealing temperatures by any of the authors. In our study, it has been observed that the highest contact angle and the lowest CAH for FAS passivated films are achieved at an annealing temperature of 523 K, as shown in Fig. 5. The CAH remains nearly constant in the temperature range of 450 K and 550 K, but increases slightly at 573 K perhaps due to the initiation of thermal desorption of FAS molecules at that temperature. At an annealing temperature of 673 K, CAH increases to a high value of 40° and water drops stick to the surface, which is probably due to the complete desorption/decomposition of FAS molecules at that temperature. Further study would be necessary to understand the phenomenon of thermal desorption and decomposition of FAS molecules with the annealing temperatures as reported for stearic acid and oleic acid molecules on metallic nanostructure surfaces [15, 16]. We have observed that FAS coated surfaces are more stable than SA coated ones against thermal annealing due to the strong oxide bonds of FAS molecules as compared to the weak hydrogen bonds of SA molecules.

4. CONCLUSIONS

Nanostructured patterns have been created on aluminum surfaces by chemical etching using diluted hydrochloric acid. FESEM images of these surfaces show a cloudy fiber-like structure showing the etched pits prominently. AFM investigation of samples etched for periods of 2 minutes and 4 minutes shows that the rms roughnesses are 165 nm and 366 nm, respectively. Larger micropits are created on rougher surfaces, resulting from prolonged etching, which allows entrapment of greater amount of air that effectively increases the contact angle of water after passivation. Passivation of these surfaces has been carried out using FAS and SA molecules which terminate the surface with $-CF_3$ and $-CH_3$ heads, respectively. A contact angle higher than 170° has been achieved on both surfaces with a contact angle hysteresis of 7.3° and 9.3° for $-CF_3$ and $-CH_3$ terminated surfaces, respectively. However, SA passivated surfaces turn hydrophilic at 573 K due to a complete thermal desorption and decomposition of the SA molecules at this temperature. FAS coated surfaces, on the other hand, show a contact angle hysteresis lower than 4° at the same temperature of 573 K and become hydrophilic at 673 K, most likely due to the thermal desorption as well as the decomposition occurring at this temperature. FAS coated surfaces are thus thermally more stable at higher temperatures than SA coated ones.

Acknowledgments

This research was carried out within the framework of the NSERC/Hydro-Quebec/ UQAC Industrial Chair on Atmospheric Icing of Power Network Equipment (CIGELE) and the Canada Research Chair on Engineering of Power Network Atmospheric Icing (INGIVRE) at Université du Québec à Chicoutimi. The authors would like to thank all the partners of CIGELE/INGIVRE for their financial support. The authors are also thankful to Ms. Hélène Grégoire, of CNRC in Chicoutimi, for providing access to FESEM facilities.

REFERENCES

1. W. Barthlott and C. Neinhuis, *Planta* **202**, 1 (1997).
2. W. Lee, M.-K. Jin, W.-C. Yoo and J.-K. Lee, *Langmuir* **20**, 7665 (2004).
3. X. Gao and L. Jiang, *Nature* **432**, 36 (2004).
4. G. S. Tzeng, H. J. Chen, Y. Y. Wang and C. C. Wan, *Surface Coating Technol.* **89**, 108 (1997).
5. H. Y. Erbil, A. L. Demirel, Y. Avci and O. Mert, *Science* **299**, 1377 (2003).
6. J. Bico, C. Marzolin and D. Quéré, *Europhys. Lett.* **47**, 220 (1999).
7. T. Onda, S. Shibuichi, N. Satoh and K. Tsujii, *Langmuir* **12**, 2125 (1996).
8. K. Tadanaga, J. Morinaga and T. Minami, *J. Sol.-Gel. Sci. Technol.* **19**, 211 (2000).
9. S. Shibuichi, T. Yamamoto, T. Onda and K. Tsujii, *J. Phys. Chem.* **100**, 19512 (1996).
10. S. Coulson, I. Woodward, J. Badyal, S. A. Brewer and C. Willis, *J. Phys. Chem. B* **104**, 8836 (2000).
11. S. Suzuki, A. Nakajima, Y. Kameshima and K. Okada, *Surface Sci.* **557**, L163 (2004).
12. B. Qian and Z. Shen, *Langmuir* **21**, 9007 (2005).
13. D. K. Sarkar and M. Farzaneh, *Proceedings of Nanotech 2006, Boston*, Vol. 3, p. 166 (2006).
14. C. R. Kessel and S. Granick, *Langmuir* **7**, 532 (1991).
15. H. J. Jeong, D. K. Kim, S. B. Lee, S. H. Kwon and K. Kadonoz, *J. Colloid. Interface. Sci.* **235**, 130 (2001).
16. H. M. Shang, Y. Wang, S. J. Limmer, T. P. Chou, K. Takahashi and G. Z. Cao, *Thin Solid Films* **472**, 37 (2005).
17. M. Futamata, X. Gaia and H. Itohb, *Vacuum* **73**, 519 (2005).
18. S. J. Lee and K. Kim, *Vibrational Spectrosc.* **18**, 187 (1998).
19. C. Consalvo, S. Panebianco, B. Pignataro, G. Compagnini and O. Puglisi, *J. Phys. Chem. B* **103**, 4687 (1999).
20. K.-W. Wang, S.-R. Chung, W.-H. Hung and T. P. Perng, *Appl. Surface Sci.* **252**, 8751 (2006).
21. V. P. Dieste, O. M. Castellini, J. N. Crain, M. A. Eriksson, A. Kirakosian, J.-L. Lin, J. L. Mc. Chesney, F. J. Himpsel, C. T. Black and C. B. Murray, *Appl. Phys. Lett.* **83**, 5053 (2003).

Contact Angle, Wettability and Adhesion, Vol. 5, pp. 279–285
Ed. K.L. Mittal
© VSP 2008

Effect of temperature on superhydrophobic zinc oxide nanotowers

N. SALEEMA, D. K. SARKAR and M. FARZANEH*

Canada Research Chair on Atmospheric Icing Engineering of Power Networks (INGIVRE) and Industrial Chair on Atmospheric Icing of Power Network Equipment (CIGELE) Icing Research Building, Université du Québec à Chicoutimi (UQAC) 555, Boulevard de l'Université, Chicoutimi, Québec G7H 2B1, Canada

Abstract—Superhydrophobic zinc oxide nanotowers have been grown successfully on silicon surfaces by chemical bath deposition (CBD). Chemically and ultrasonically cleaned silicon substrates were immersed in a beaker containing 100 ml of aqueous 0.1 M $Zn(NO_3)_2$ and 4 ml of 28% aqueous NH_4OH solution. The CBD was performed in an oven at 70°C for 20 minutes. The dried samples were passivated using stearic acid (SA) for lowering the surface energy. The SA-passivated samples were annealed at various temperatures ranging from 70°C to 350°C. Water contact angle measurements and Fourier transform infrared (FTIR) spectroscopy investigation were performed on the samples at each annealing temperature. The contact angle of water at various temperatures remains nearly constant, at 160°, with a hysteresis of less than 5° and the tendency for the water drop to roll off easily from the surface. FTIR studies reveal that the amount of SA starts to decrease above 200°C and is completely lost at 350°C, leaving the sample highly hydrophilic. The partial passivation of the surface due to annealing and the consequent behavior of water contact angle is discussed.

Keywords: Superhydrophobicity; ZnO nanotowers; stearic acid; thermal desorption.

1. INTRODUCTION

It is well known that surface wettability is a function of surface roughness, and surfaces for which wetting is nearly zero are said to exhibit superhydrophobicity. Superhydrophobicity or non-wettability is commonly encountered in living objects such as lotus leaves [1]. Such a property is termed the "Lotus effect". Recently, hydrophobic and superhydrophobic surfaces have usually been produced in two steps involving surface geometry and chemistry [2]. Creating a rough structure on a surface, followed by a chemical modification with compounds with low surface free energy, such as fluorinated or silicone compounds, results in superhydrophobicity. Recent studies also indicate the necessity of a two tier surface roughness to achieve a very high superhydrophobicity [3, 4]. The Wenzel and Cassie-Baxter models

---

*To whom correspondence should be addressed. Tel.: 1-418-545-5011 ext. 5044; Fax: 1-418-545-5032; e-mail: farzaneh@uqac.ca

explain the impact of roughness on surface wettability [5, 6]. Superhydrophobic nanostructured surfaces can have important applications as coatings for cables, insulators, glass windows, windshields, lenses, etc.

Zinc oxide (ZnO) has recently attracted attention in the field of nanoscience for its unique structural, optical and electrical properties [7–9]. There have been several recent attempts to create several kinds of ZnO nanostructures fabricated by various sophisticated procedures [10, 11]. Among the techniques used, chemical bath deposition (CBD) has achieved recognition for its simplicity, low temperature and low cost process [12]. Although superhydrophobicity on these CBD ZnO films has been achieved by the use of organic compounds, such as fluoroalkylsilanes (FAS), the strength of bonding of these compounds to the ZnO has not yet been described [13]. Recently, studies on the thermal decomposition of surfactant coatings, such as oleic acid, stearic acid and poly(ethylene glycol), have been carried out on nanostructured metal surfaces [14, 15], although there are no reports on the temperature effect of these organic molecules on oxides.

In this paper, we present a novel highly superhydrophobic ZnO nanostructure, the 'nanotower', grown on silicon substrates prepared by the CBD method, which shows a rough binary structure composed of uniform nanosteps throughout each nanotower. We report a remarkable superhydrophobicity of these ZnO nanotowers achieved by passivating them using stearic acid, and we demonstrate the effect of temperature on the superhydrophobicity and bonding of the stearic acid to the ZnO nanotowers.

2. EXPERIMENTAL DETAILS

The aqueous solution used for the CBD consisted of 100 ml of aqueous 0.1 M $Zn(NO_3)_2$ and 4 ml of 28% aqueous NH_4OH solution. Silicon substrates, $1'' \times 1''$, were cleaned ultrasonically successively with acetone, propanol and distilled water for 10 minutes each at an ultrasonic solution temperature of 22°C. The silicon substrates were immersed in the aqueous solution for coating deposition. The beaker containing the solution was placed in an oven at 70°C and the deposition was carried out for 20 minutes. These samples were rinsed in a beaker containing methanol to remove any loose particles, dried in an oven at 70°C for several hours, cooled to room temperature, and then passivated with 2×10^{-3} M stearic acid in acetone. The passivated samples were annealed at different temperatures ranging from 70°C to 350°C, and tested for superhydrophobicity using a contact angle goniometer (Krüss GmbH, Germany) after each annealing temperature. FTIR (PerkinElmer Spectrum One) studies were performed after each annealing temperature to analyze the bonding of the SA molecules to the ZnO surfaces. The microstructural investigations of these samples were examined by field emission scanning electron microscopy (FEGSEM, Leo 1525) and atomic force microscopy (AFM) (Digital Nanoscope IIIa by Digital Instruments).

3. RESULTS AND DISCUSSION

The ZnO nanotower deposition process involves chemical reaction between zinc nitrate and ammonium hydroxide to give ZnO [12]. The deposited ZnO nanotowers have a hexagonal pattern with several nanosteps of uniform step-size throughout the top few hundred nanometers, as seen in the FESEM images in Fig. 1. While Fig. 1(a) shows the presence of randomly oriented ZnO nanotowers, the hexagonal pattern and the presence of nanosteps on the nanotowers are seen in Fig. 1(b). The height of these nanotowers is ∼700 nm, with a tower width of ∼500 nm. This observation is complemented by the AFM results in Fig. 2. Figure 2(a) is a two-dimensional view of the ZnO nanotowers captured in the tapping mode with a scan size of 5 μm ×5 μm. Cross-sectional analysis (Fig. 2(b)) shows that the width of the nanotowers is roughly 500 nm and the step-size is ∼25–

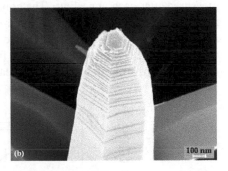

Figure 1. FESEM images of ZnO nanotowers grown on glass substrates at (a) low magnification; inset shows the image of a water drop, (b) ZnO nanotowers at high magnification.

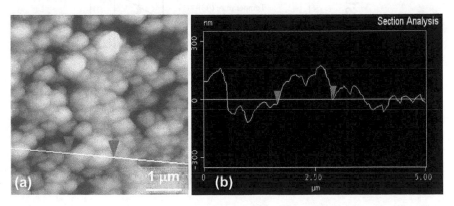

Figure 2. (a) 2-D view of tapping mode AFM image, (b) section analysis of ZnO nanotowers on a scan size 5 μm ×5 μm.

30 nm, both horizontally and vertically with a root mean square (rms) roughness of ∼50 nm. These observations reveal the presence of a rough binary structure with a combination of nanosteps on the nanotowers, which is a basic geometrical requirement to achieve superhydrophobicity after further chemical modification [2].

The superhydrophobicity of these ZnO nanotowers was evaluated by passivating them using stearic acid and drying them at 70°C in an oven for 30 minutes. The contact angle of water achieved on these nanotowers is as high as ∼173° with a contact angle hysteresis as low as ∼1.3°, the water drops roll off the surface even with the slightest tilt. This high superhydrophobicity indicates that the hydrophilic tail (-COOH) of stearic acid ($CH_3(CH_2)_{16}COOH$) reacts with the ZnO during passivation while their hydrophobic heads (-CH_3) remain upwards.

The ZnO nanotowers were annealed in air at various temperatures from 70–350°C to study the bonding of stearic acid to the surface, and the water contact angle was measured after each annealing. The contact angle, as shown in the inset of Fig. 3, remains >170° and shows only a minor decrease up to 250°C at which the contact angle is still >160°. The contact angle hysteresis (Fig. 3) undergoes an increase with increasing annealing temperature, from 1.3° at 70°C to 6° at 250°C, although the water drops still roll off the surface easily. However, the contact angle continues

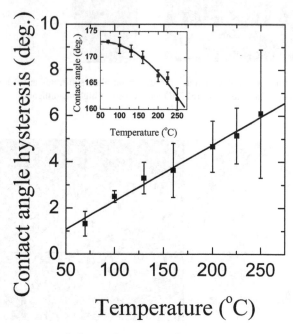

Figure 3. Water contact angle hysteresis vs. annealing temperature, Inset: Contact angle vs. annealing temperature.

to fall with increasing annealing temperature and at 350°C it was no longer possible to measure it, as the water drop spread completely on the surface.

Figure 4 shows the FTIR spectra, in the range of 2550–3150 cm^{-1}, acquired at each annealing temperature. The spectra have been shifted upwards for easy comparison showing only the -CH$_n$ peaks of stearic acid. The two peaks at 2919 cm^{-1} and 2850 cm^{-1} belong, respectably, to the asymmetric and symmetric C-H stretching modes of the -CH$_2$ group of stearic acid with a peak at 2958 cm^{-1} appearing due to the asymmetric in-plane C-H stretching mode of the -CH$_3$ group [16] and the ZnO peak was observed at 418 cm^{-1}(not shown in the figure). It is clear from the spectra that the two stearic acid peaks remain nearly unchanged until 160°C. There is a drastic change at 200°C in the intensity of these peaks which decreases with increasing temperature. The contact angle and contact angle hysteresis results have a close correlation with the FTIR results in a way that the nearly constant contact angle with a very low contact angle hysteresis up to 160°C is due to the presence of nearly constant stearic acid density on the ZnO nanotowers. As the density of the stearic acid starts to decrease near 200°C, the contact angle begins to decrease, with increasing, although not appreciable, contact angle hysteresis. At 350°C, the water drop spreads completely on the surface and the FTIR spectrum shows a zero intensity of the -CH$_n$ peak. Pérez-Dieste *et al.* [14] studied the thermal decomposition of oleic acid on Co and Ni nanocrystals, and found that the desorption of these molecules began at ~200°C and dehydrogenated at ~400°C. In the case of our stearic acid/ZnO system, according to the FTIR

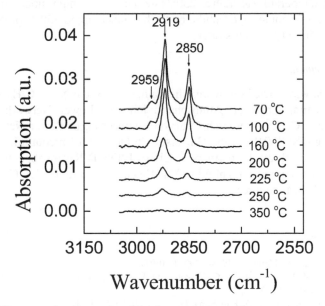

Figure 4. FTIR spectra showing stearic acid peaks at various annealing temperatures.

analysis, the intensity of the FTIR peaks drastically reduces at 200°C due to the thermal desorption of stearic acid. Although, Pérez-Dieste *et al.* [14] studied oleic acid/metal system, the results are in agreement with our stearic acid/ZnO system. However, in the case of Pérez-Dieste *et al.* [14], dehydrogenation takes place at 400°C which in our case is probably at 350°C and the sample becomes completely hydrophilic, probably due to the presence of carbon on the surface as observed by those authors.

4. CONCLUSION

Highly superhydrophobic ZnO nanotowers have been grown on silicon substrates by the chemical bath deposition method. These nanotowers on passivation by stearic acid give a water contact angle as high as ~173°, with a contact angle hysteresis as low as 1.3°. The effect of temperature on these nanotowers has been studied by annealing them in air at different temperatures, and the FTIR spectra and contact angle and contact angle hysteresis measurements carried out at each annealing temperature indicate that the density of stearic acid starts to drastically decrease from 200°C, at which the contact angle reduces to ~166° with a contact angle hysteresis increase to 4.6°. At 350°C, with zero intensity of the FTIR peaks of stearic acid, the water drop spreads on the surface making it completely hydrophilic. These observations illustrate that there exists a strong bonding of the stearic acid to ZnO even at temperature as high as 250°C showing a very high superhydrophobicity and the surface becomes hydrophilic by further annealing to 350°C probably due to dehydrogenation at this temperature.

Acknowledgements

This research was carried out within the framework of the NSERC/Hydro-Quebec/ UQAC Industrial Chair on Atmospheric Icing of Power Network Equipment (CIGELE) and Canada Research Chair on Engineering of Power Network Atmospheric Icing (INGIVRE) at the Université du Québec á Chicoutimi. The authors would like to thank all the partners of CIGELE/INGIVRE for their financial support. The authors are also thankful to Ms. Hélène Grégoire, CNRC Chicoutimi for providing FESEM facilities.

REFERENCES

1. W. Barthlott and C. Neinhuis, *Planta* **202**, 1 (1997).
2. Y. H. Yang, Z. Y. Li, B. Wang, C. X. Wang, D. H. Chen and G. W. Yang, *J. Phys.: Condens. Matter.* **17**, 5441 (2005).
3. X.-T. Zhang, O. Sato and A. Fujishima, *Langmuir* **20**, 6065 (2004).
4. D. M. Soolaman and H. Z. Yu, *J. Phys. Chem. B* **109**, 17967 (2005).
5. R. N. Wenzel, *Ind. Eng. Chem.* **28**, 988 (1936).

6. A. Cassie and S. Baxter, *Trans. Faraday Soc.* **40**, 546 (1944).
7. R. Li, S. Yabe, M. Yamashita, S. Momose, S. Yoshida, S. Yin and T. Sato, *Mater. Chem. Phys.* **75**, 39 (2002).
8. S. Liang, H. Sheng, Y. Liu, Z. Hio, Y. Lu and H. Shen, *J. Cryst. Growth.* **225**, 110 (2001).
9. N. Saito, H. Haneda, T. Sekiguchi, N. Ohashi, I. Sakaguchi and K. Koumoto, *Adv. Mater.* **14**, 418 (2002).
10. X. Y. Kong, Y. Ding, R. Yang and Z. L. Wang, *Science* **303**, 1348 (2004).
11. S. Yin and T. Sato, *J. Mater. Chem.* **15**, 4584 (2005).
12. N. Saleema, D. K. Sarkar, M. Farzaneh and E. Sacher, *Proceedings of Nanotech 2006, Boston*, pp. 158-161 (2006).
13. M. Li, J. Zhai, H. Liu, Y. Song, L. Jiang and D. Zhu, *J. Phys. Chem. B* **107**, 9954 (2003).
14. V. Pérez-Dieste, O. M. Castellini, J. N. Crain, M. A. Eriksson, A. Kirakosian, J.-L. Lin, J. L. McChesney, F. J. Himpsel, C. T. Black and C. B. Murray, *Appl. Phys. Lett.* **83**, 5053 (2003).
15. K.-W. Wang, S.-R. Chung, W.-H. Hung and T. P. Perng, *Appl. Surf. Sci.* **252**, 8751 (2006).
16. A. Mills, S.-K. Lee, A. Lepre, I. P. Parkin and S. A. O'Neill, *Photochem. Photobiol. Sci.* **1**, 865 (2002).

Contact Angle, Wettability and Adhesion, Vol. 5, pp. 287–293
Ed. K.L. Mittal
© VSP 2008

Superhydrophobic properties of silver coated copper

A. SAFAEE, D. K. SARKAR and M. FARZANEH*

Canada Research Chair on Atmospheric Icing Engineering of Power Networks (INGIVRE) and Industrial Chair on Atmospheric Icing of Power Network Equipment (CIGELE) Icing Research Building, Université du Québec à Chicoutimi (UQAC) 555, Boulevard de l'Université, Chicoutimi, Québec G7H 2B1, Canada

Abstract—Superhydrophobic properties have been studied for silver nanoparticles covered copper surfaces. The silver nanoparticles have been obtained by reducing silver ions from silver nitrate solution onto copper substrates by means of a galvanic exchange reaction. The size of the nanoparticles has been controlled by varying the concentration of the solution. The contact angle and contact angle hysteresis have been studied on these samples after passivating with low surface energy stearic acid organic molecules. The obtained water contact angle is as high as 156° with a hysteresis as low as 4° for one minute coating using 24.75 mM solution. However, a strong dependence of contact angle and hysteresis on the silver ion concentration in the solution has been observed.

Keywords: Superhydrophobicity; galvanic exchange reaction; silver nanostructure; contact angle; scanning electron microscopy.

1. INTRODUCTION

Hydrophobicity is currently the focus of considerable research. Wettability is an important property of solids and is controllable by both the chemical composition and the geometrical structure of the surfaces [1, 2]. Many plants and insects in nature such as water striders, lotus leaves, etc., exhibit superhydrophobicity due to the presence of the specific microstructure on their surfaces [3]. Since the interactions between solids and water drops are limited to the outermost layer of the surface, it is possible that the superhydrophobic properties exhibited by these plants and insects are due to the presence of an outer shell such as a waxy coating over their rough microstructure [4]. Contact angle of water on such surfaces is usually larger than 150° and water drops on these surfaces roll off easily even with the slightest angle of inclination of the surface. Rolling drops remove contamination from the surfaces and this property is referred to as 'self-cleaning'. Accordingly, various phenomena such as the adhesion of snow or raindrops, oxidation, and friction drag are expected to be inhibited or reduced on such surfaces [5].

*To whom correspondence should be addressed. Tel.: 1-418-545-5011 ext. 5044; Fax: 1-418-545-5032; e-mail: farzaneh@uqac.ca

In recent years, to control the structure, dimensions, and regularity of the surface patterns, many strategies have been used such as photolithography [6], plasma etching [7], chemical etching [8, 9], sol-gel [10], etc., in order to achieve superhydrophobicity. Recently, electrodeposition of silver aggregates has been performed on copper surfaces [11] to obtain superhydrophobic surfaces. The galvanic exchange reaction method has been recently used by Song *et al.* [12] to create silver nanostructures on silicon surfaces, but no results on its superhydrophobicity were reported.

In this paper, we report on the achievement of superhydrophobicity on silver nanostructures deposited on copper substrates using a galvanic exchange reaction. Also, we discuss the superhydrophobic behavior of these surfaces, as achieved by passivation with stearic acid organic molecules using contact angle and contact angle hysteresis measurements and scanning electron microscopy.

2. EXPERIMENTAL

One-square-inch copper substrates were ultrasonically cleaned with a 0.1 M sodium hydroxide solution for 10 minutes, followed by rinsing with water. Some of the copper substrates were etched with dilute nitric acid with varying concentration (0–50% volume ratio) for 2 minutes. The silver particles were deposited by immersing the copper substrates in silver nitrate solutions for one minute. The solution concentration was varied from 0 to 396 mM. The etched copper substrates were coated using a 24.75 mM silver nitrate solution. The silver coated samples were rinsed in water and dried at 60°C for 30 minutes. The dried samples were passivated with a 2 mM solution of stearic acid in acetone for 20 minutes. The passivated samples were dried at 60°C for 30 minutes. The surface morphology of the samples was investigated by field emission scanning electron microscopy (FEGSEM, Leo 1525) and equilibrium contact angle measurements were made using a Krüss DSA100 drop shape analyzer. A standard and very commonly-used procedure as reported in the literature [13] was followed to measure the contact angle hysteresis, i.e., the difference between the advancing and receding contact angles. In this method, water drops of volume ∼5 μL were suspended from a needle and brought into contact with the sample surfaces using a computer controlled device provided by Krüss. The advancing and receding contact angles were measured by holding the water drop with stationary needle in contact with the surface and moving the contact angle goniometer stage in one direction. The contact angle data were acquired by fitting the symmetric water drops using the Laplace-Young equation and the advancing and receding contact angles were measured by fitting the asymmetric water drops using the tangent-2 method [14]. Each numerical result is an average of the data from 7 points on each sample.

3. RESULTS AND DISCUSSION

Silver ions reduce to silver atoms in the presence of copper due to the electrochemical potential difference between silver and copper. This ion reduction method has been employed to deposit silver coating on copper surfaces using silver nitrate solution. These samples were passivated with stearic acid. It is reported that a stearic acid molecule is chemisorbed on silver as carboxylate with its two oxygen atoms bound symmetrically to the surface [15]. Figure 1 shows the water contact angle and the contact angle hysteresis for the samples prepared with different silver nitrate concentrations. By raising the initial concentration from 13.2 mM to 24.75 mM, Fig. 1(a) shows an increment of contact angle from 137° to 156°, while Fig. 1(b) depicts a remarkable decrement in contact angle hysteresis from 29.5° to 3.6°. Increasing the solution concentration further to 49.5 mM provides a reduced contact angle of 147° with an increased hysteresis of 4.9°. The contact angle further reduces to 141° increasing the hysteresis to 15.5° at an increased concentration of 396 mM. These observations indicate that there is a critical point in the concentration level of the silver nitrate solution used up to which the contact angle increases with a decreasing hysteresis. After this critical point, the contact angle starts to decrease while hysteresis increases.

SEM analyses of the samples prepared with three different solution concentrations (13.2, 24.75 and 396 mM) were performed to study the contact angle behavior at the lowest, intermediate and highest concentrations used (Fig. 2). It is clear from Fig. 2 that not only the size of the fractal structure of silver increases with increasing silver ion concentration, but also the number of noticeable voids. These images reveal that the volume of the voids increases with the concentration. At lower concentration, a more compact coating has been achieved due to slower reaction. Larger fractal flakes are created at higher concentration due to faster reaction, thus increasing the volume of the voids. Figure 2(a) shows that the sizes of the voids are almost indistinguishable for the coating deposited from a 13.2 mM solution. However, the sizes of the voids are ~5 μm and ~10 μm for the samples prepared using 24.75 mM and 396 mM silver nitrate solutions, as shown in Figs. 2(b) and 2(c), respectively. When the void size is relatively small, as with the 13.2 mM concentration, the surface is smooth and a lower contact angle and a higher hysteresis are obtained. When the void size is very large, as is the case for the sample prepared using a 396 mM concentration, the contact angle is again low with increased hysteresis. However, the highest contact angle with lowest hysteresis was achieved with an intermediate void size using a 24.75 mM solution. Both the contact angle and SEM analyses indicate that the critical value for achieving superhydrophobicity is at a 24.75 mM solution concentration. In the literature, galvanic exchange reaction has been used to create fractal-like gold nanostructures on silicon surfaces that showed superhydrophobic behavior after passivation with n-dodecanethiol [16]. In the present case, stearic acid passivated silver nanostructures on copper surfaces prepared from a 24.75 mM solution for one minute provides a very high contact angle of ~156°. On the other hand, similar results were obtained on n-dodecanethiol

290 A. *Safaee* et al.

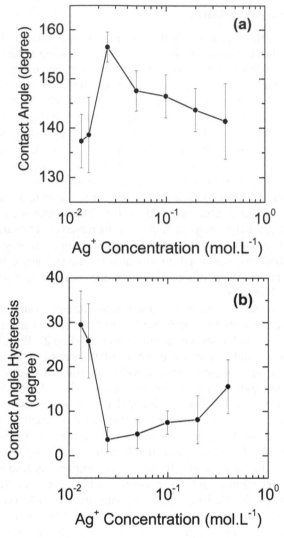

Figure 1. (a) Water contact angle and (b) contact angle hysteresis of silver coated copper samples for various Ag[+] ion concentrations.

passivated gold nanostructures on a silicon surface prepared from a 25 mM solution for 40 minutes [15]. This difference in time is probably due to the higher reaction rate of silver on conducting copper than that of gold on semiconducting silicon.

The intrinsic electrochemical potential difference between copper and hydrogen makes it possible to use the galvanic exchange reaction to etch the copper samples. Figure 3 (curve a) shows the contact angle results for etched copper surfaces

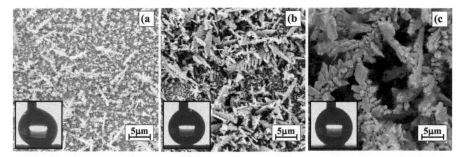

Figure 2. SEM images of copper samples coated with silver particles using Ag^+ concentrations of (a) 13.2, (b) 24.75 and (c) 396 mM.

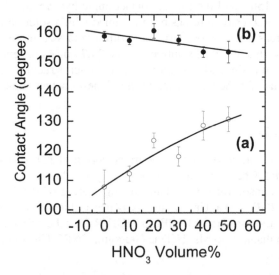

Figure 3. Water contact angle on (a) uncoated and (b) silver coated copper substrates as a function of nitric acid concentration. All the substrates were etched with nitric acid for 2 minutes.

passivated using stearic acid. Figure 3 (curve b) shows the contact angles on silver coated samples using a 24.75 mM silver nitrate solution after etching the substrates as in Fig. 3 (curve a). The contact angle is 108° for unetched and uncoated samples which has been enhanced to 160° after being coated with silver. Although the contact angle increased to 130° after etching with a 50% acid concentration (Fig. 3 (curve a)), it decreased to 153° on the etched and silver coated surface (Fig. 3 (curve b)). After etching with a 20% acid solution, the contact angle obtained is 123° which further increases to 161° due to silver coating. From Fig. 3 (curve a), it is clear that the contact angle increases with increasing the concentration of etchant. Usually, etching begins at points of dislocation on the surface and continues both parallel and perpendicular to the surface. The number of etch pits depends on the concentration of the etchant used. Thus, by increasing the etchant concentration, more pits are created, increasing the surface roughness and, consequently, the contact angle. On the other hand, a slow decrease of contact angle has been observed on these samples after silver coating.

4. CONCLUSION

The galvanic exchange reaction has been used to create nanostructured silver films on copper surfaces. Although, the concentration of silver ions was varied from 0 to 396 mM, its critical value was found to be 24.75 mM, at which the highest contact angle, ~156°, and the lowest contact angle hysteresis, 4°, were achieved. A very compact microstructure has been obtained with a solution concentration of 13.2 mM, whereas a very loose structure with a void size more than 10 μm has been observed with a 396 mM solution. At the 24.75 mM concentration, however, the presence of an optimized ~5 μm void in the microstructure of the silver films might be the cause for the occurrence of the highest contact angle with the lowest hysteresis.

Acknowledgements

This research was carried out within the framework of the NSERC/Hydro-Quebec/ UQAC Industrial Chair on Atmospheric Icing of Power Network Equipment (CIGELE) and Canada Research Chair on Engineering of Power Network Atmospheric Icing (INGIVRE) at Université du Québec à Chicoutimi. The authors would like to thank all the partners of CIGELE/INGIVRE for their financial support. The authors are also thankful to Ms. Hélène Grégoire, CNRC Chicoutimi for providing FESEM facilities.

REFERENCES

1. L. Feng, S. Li, Y. Li, H. Li, L. Zhang, J. Zhai, Y. Song, B. Liu, L. Jiang and D. *Zhu, Adv. Mater.*, **14**, 1857 (2002).
2. Z. Z. Gu, H. Uetsuka, K. Takahashi, R. Nakajima, H. Onishi, A. Fujishima and O. Sato, *Angew. Chem. Int. Ed.*, **42**, 894 (2003).
3. W. Barthlott and C. Neinhuis, *Planta*, **202**, 1 (1997).
4. S. A. Kulinich and M. Farzaneh, *Vacuum*, **79**, 255 (2005).
5. T. Kako, A. Nakajima, H. Irie, Z. Kato, Z. Uematsu and T. Watanabe, *J. Mater. Sci.*, **39**, 547 (2004).
6. D. Oner and T. J. McCarthy, *Langmuir*, **16**, 7777 (2000).
7. R. Olde, J. G. A. Terlingen, G. H. M. Engbers and J. Feijen, *Langmuir*, **15**, 4847 (1999).
8. N. J. Shirtcliffe, G. McHale, M. I. Newton and C. C. Perry, *Langmuir*, **21**, 937 (2005).
9. B. Qian and Z. Shen, *Langmuir*, **21**, 9007 (2005).

10. K. Satoh and H. Nakazumi, *J. Sol–Gel Sci. Technol.*, **27**, 327 (2003).
11. N. Zhao, F. Shi, Z. Wang and X. Zhang, *Langmuir*, **21**, 4713 (2005).
12. Y. Y. Song, Z. D. Gao, J. J. Kelly and X. H. Xia, *Electrochem. Solid-State Lett.*, **8**, C148 (2005).
13. M. Callies, Y. Chen, F. Marty, A. Pépin and D. Quéré, *Microelectron. Eng.*, **100**, 77 (2005).
14. Krüss GmbH, DSA1 v1.9-03 (User Manual).
15. S. J. Lee and K. Kim, *Vib. Spectrosc.*, **18**, 187 (1998).
16. C. Wang, Y. Song, J. Zhao and X. Xia, *Surf. Sci.*, **600**, L38 (2006).

Contact Angle, Wettability and Adhesion, Vol. 5, pp. 295–308
Ed. K.L. Mittal
© VSP 2008

Surfactants adsorption at hydrophobic and superhydrophobic solid surfaces

MICHELE FERRARI *

CNR-National Research Council – IENI, Institute for Energetics and Interphases, via De Marini 6, 16149 Genoa, Italy

Abstract—The adsorption properties of surfactant molecules at solid surfaces with different degrees of hydrophobicity have been reviewed together with original research data in case of superhydrophobic surface coatings, with a water contact angle (CA) close to or greater than 150°, in both liquid-air and liquid-liquid systems.

The studies of the adsorption properties of amphiphiles on water repellent solid surfaces, especially in conditions of extreme hydrophobicity, have recently attracted great interest both from fundamental and applied points of view.

The nature of the amphiphilic molecules and the substrate surface topography also play important roles in research and technological fields where the spreading control of the amphiphiles solutions is influenced by the distribution properties of the surfactant between two immiscible phases.

Keywords: Surfactants; adsorption; superhydrophobicity; liquid-liquid systems.

1. INTRODUCTION

Surfactants are widely used in many technological fields involving liquid–solid interfaces, such as oil recovery, flotation, colloidal systems, wetting phenomenon, or self-assembly properties which are just some examples how the specifically tailored features of the interface require an understanding of the structure and morphology of the adsorbed layers.

The behaviour of surfactant molecules adsorbed at hydrophobic solid surfaces has been the subject of a wide literature while, but in case of extreme water repellence, or superhydrophobicity, reference data are still scarce, despite of its potential applications interest.

Surfaces known as superhydrophobic or ultrahydrophobic, showing a water contact angle greater than 150°, have recently attracted great interest not only in fundamental research, but also a wide range of different practical applications is offered in fact by superhydrophobic coating films ranging, to name a few, from

---

*Tel.: +39 010 6475723; Fax: +39 010 6475700; e-mail: m.ferrari@ge.ieni.cnr.it

corrosion resistant films to protective layers for solar panels, cars, buildings glasses and sunglasses and electronic components.

For example, Bico et al. [1] showed that on such surfaces with a small area in contact with water, a significant decrease of resistance to liquid drainage can be expected as well as inhibition of various related phenomena involving physico-chemical interactions in an aqueous environment like adhesion and oxidation. Öner and McCarthy [2] also found these surfaces to have low surface energy and high roughness at nanometric scale.

It has been known for a long time that an increase of surface roughness of a hydrophobic solid substrate often results in a substantial increase in the degree of hydrophobicity. The key role played by surface roughness as an amplifier of the hydrophobicity in natural and artificial substrates by combining geometrical and chemical characteristics was shown by Feng et al. [3]. Nevertheless, although these surfaces are generally fabricated by combining appropriate surface roughness with low surface energy materials, it has been found by Feng et al. [4] that such effect can also be attained starting from an amphiphilic polymer coating.

Zhang et al. [5] proposed some simple methods to obtain a superhydrophobic film by mechanical extension of a Teflon film. By this technique the CA of a drop of water effectively increased from 118 to 165 degrees by an extension of ca. 190%.

More recently these features, coupled with the modern self-assembly and micro-fabrication techniques, have been exploited in fields where such extremely high CAs and very low flow resistance are required. In the particular field of biomimetics, us-ing a galvanic exchange reaction, Wang et al. [6] fabricated stable gold coating exploiting the cooperative effect of microstructures and nanostructures with various morphologies under different conditions, such as concentration of gold salt, light, temperature, ultrasonication, or addition of surfactants. After modification of the films by immersion in n-dodecanethiol (ethanol solution), a superhydrophobic sur-face (water CA 165°), similar to the lotus leaf, was obtained.

Surface undulation has been also taken into account by Taniguchi et al. [7] in a quantitative way regarding the effect of rough surfaces on the wetting of porous substrates. As reported by Patankar and coworkers [8–10] double roughness composed of a rough base together with smooth pillars prevents a surface from wearing, showing that double (or multiple) roughness structures or slender pillars are appropriate surface geometries to develop "self-cleaning" surfaces. The key motivation behind the double-structure roughness is to mimic the microstructure of the extremely hydrophobic surface of plant leaves (such as lotus). The calculation procedure presented was used to obtain optimal surface geometries to fabricate self-cleaning surfaces. The analysis could also be generalized for a fractal surface or multiple-roughness structures.

2. SURFACTANTS ADSORPTION AT HYDROPHOBIC SURFACES

According to van Oss *et al.* [11], one of the most hydrophobic surfaces known is the air side of the water-air interface and even more hydrophobic than the surfaces of nonpolar condensed-phase compounds or materials such as hydrocarbons or Teflon. The water-air interface hydrophobicity is considered the main cause of the large increase in CA of water drops on rough surfaces of apolar materials, as compared with the water CA on smooth surfaces of the same materials. A very porous fractal surface, composed of a solid surface fraction of few percents and air, can support a water drop reaching CAs higher than 170°, very close to the theoretical maximum of 180°. Apolar molecules and the apolar side of amphiphilic molecules (such as surfactant hydrocarbon chain) are attracted by the water-air interface due to hydrophobic interactions.

Adsorption on solid interfaces is often characterized by self-assembly of the amphiphilic molecules [12, 13] and most of these properties are in fact expression of the nanoscopic features of the adsorbed layer.

Self-assembly phenomena are influenced by physico-chemical conditions such as bulk concentration and temperature, which affect the shape and the periodic structure of the adsorbed layer. In the bulk, surfactant molecules associate into aggregates, such as micelles or vesicles, above a critical micelle concentration (cmc). Similarly, self-assembly at interfaces occurs beyond a critical surface aggregation concentration (csac) [14].

Systematic studies have been performed with the aim to investigate the relationship between surface-aggregate structure and the nature of the surfactant and the interface and thus to understand the aggregate shape in terms of intermolecular forces [15, 16].

Amphiphilic molecules organise at the interface in periodic stripes as predominant structure, and, in agreement with the data reported in the literature [17, 18], the nature of the hydrophilic head is found to play a key role in influencing periodicity and morphology of the stripes. Different authors [19, 20] have reported these stripes to be molecular structures formed on top of a horizontal bilayer of surfactant molecules arranged in a head-to-head configuration and directly "grafted" onto the substrate.

The influence of the chain length on the morphology of ordered stripe-like patterns adsorbed onto a hydrophobic substrate like HOPG (Highly Oriented Pyrolitic Graphite) has been investigated in a previous work by Ferrari *et al.* [21] (Fig. 1).

The subject of the investigation was the comparison of surfactants ranging from nonionic, nonionic semipolar to ionic types with the same hydrophobic chain length but different headgroups, adsorbing at a hydrophobic solid surface from a solution. Through AFM image analysis, the height and periodicity of these stripes were measured and compared with the molecular length calculated by molecular simulation software (Cerius–Mopac) and a good agreement with previously reported data [22–24], was found.

298 M. Ferrari

Figure 1. AFM image in liquid of $C_{12}DMPO$ at supramicellar concentration on HOPG graphite.

Further studies concerning the role played by the physico-chemical conditions, such as temperature and concentration, will give a more general description of the mechanism underlying these phenomena.

More recently, a novel method was found by Tsai *et al.* [25] by coupling the Langmuir-Blodgett (LB) deposition of silica particles and the formation of a self-assembled monolayer (SAM) of an alkylsilane with the aim to fabricate hydrophobic surfaces by providing the substrate with both surface roughness and low surface energy. The hydrophobic-hydrophilic balance of the silica particle surface was controlled by the adsorption of surfactant molecules and monolayers consisting of hexagonally close-packed arrays of particles on a glass substrate deposited from a Langmuir trough. A particulate film with a roughness factor of 1.9 was sintered and then hydrophobized by surface silanization. Static and dynamic water CA measurements were performed to study the effect of particle size and particle layer number on wetting properties of the particulate films. From these results the particulate films produced a static CA around 130°, independent of the size of the particles or number of layers deposited. Advancing and receding CAs in a range 150°–110° were observed with a hysteresis of about 40°, confirming the assumptions about the enhancement of the hydrophobic character by the roughness and allowing a method for switching from of the Wenzel [26] model to the Cassie-Baxter [27] regime to be designed.

Some issues arise when surfactants are involved in precipitation phenomenon at solid surfaces. Balasuwatthi *et al.* [28] investigated the wetting properties of a saturated aqueous surfactant solution on surfactant precipitates, and found that sodium and calcium salts of alkyl sulphates showed advancing CAs higher than those of alkyl trimethylammonium bromides without significant variations with surfactant/counterion ratios; therefore the wettability was not strongly affected by water hardness. In the pH range 4 to 10 fatty acid (C_{12} and C_{16}) solutions did not show any dependence of the CAs. The introduction of a second surfactant like sodium dodecyl sulphate (SDS) led to a decrease of CAs of saturated calcium dodecanoate (CaC_{12}) solutions with increasing SDS concentrations up to the critical micelle concentration of the surfactant mixture. Thus the presence of a second surfactant is effective as a wetting agent in a saturated surfactant system. The role played by the solid-liquid interfacial tension and its effect coupled with the adsorption at liquid-air interface in increasing the wetting properties was interpreted by the Young's equation. Furthermore, it was found that, by applying the Zisman method to CaC_{12}, the critical surface tension obtained was comparable to difluoroethene.

The aim to improve the wetting properties toward a superspreading action has been fulfilled in a paper [29] where the reported data deal with the mutual interactions between polyoxycthylenated surfactants (CiEjs) and comparing this behaviour to a known superwetting agent. Experiments were carried out by measuring surface tension and CA of CiEj surfactants solutions mixed with 1-dodecanol on hydrophobic surfaces. To investigate the surfactant adsorption kinetics in a dynamic way, in-situ infrared internal reflection spectroscopy and sum-frequency generation spectroscopy have been used, in order also to better understand the interfacial water structure evolution at the solid-liquid interface. Although influenced by the surfactant combination, effectively lowering the surface tension, the wetting properties were enhanced by those systems characterized by extended liquid crystalline phases as studied by cryo-TEM, cross-polarized microscopy and light scattering. These experiments focused on the surfactant adsorption at the hydrophobic solid-liquid interface, allowing the superspreading to be better interpreted at a molecular level.

Surfactant adsorption on hydrophobic surfaces has also been studied by Puttharak *et al.* [30]. In their work, the adsorption and CA measurements of three representative surfactants like sodium octyl benzene sulfonate (NaOBS), cetylpyridinium chloride (CPC) and polyoxyethylene octyl phenyl ether (OPE10) on high density polyethylene (HDPE), polystyrene (PS) and polycarbonate (PC) were conducted as a function of surfactant concentration and salinity. The results show that the adsorption of surfactant increases with increasing surfactant concentration and the surfactant adsorption reduces not only the liquid/vapor surface tension (LV), but also the solid/liquid interfacial tension (SL) resulting in a lower value of CA. The increasing polarity of the solids enhances the difference between the nature of these interfaces and an inhibition of the wetting efficiency by the effect of electrolytes like NaCl occurs, avoiding further CPC adsorption. Such an effect was not observed for NaOBS, probably due to shorter hydrophobic chain length.

3. SURFACTANT ADSORPTION AT SUPERHYDROPHOBIC SURFACES

Different methods have been proposed to enhance the water repellence by manipulating the geometry or the chemistry of the coatings [31–35]. The models of Wenzel and Cassie-Baxter are usually utilized for the interpretation of the roughness effect on the wettability properties of a solid surface.

In the Wenzel approach the space between the protrusions on the surface is assumed to be filled by the liquid and the apparent CA (θ') and the equilibrium one (θ) are then linked by

$$\cos \theta' = r \cos \theta \tag{1}$$

where r is the ratio between the true surface area and its horizontal projected area. This regime provides hydrophobic surfaces with CAs below 120°; thus, it cannot give rise to superhydrophobicity.

Superhydrophobicity can be interpreted under the Cassie-Baxter approach, suggesting that air is entrapped in the hollow spaces of the rough surface with the link between the CAs given by

$$\cos \theta' = f_1 \cos \theta - f_2 \tag{2}$$

where f_1 is the fraction of liquid area in contact with the solid and f_2 is the fraction of liquid area in contact with the trapped air.

The CA hystheresis, H, has been used to discriminate the two states: systems with larger H values are regarded belonging to the Wenzel model, where the liquid fills the grooves with adhesion to the walls; while smaller H values are attributed to the second case, where the surface is seen as composed by pillars entrapping air and strongly decreasing the space available for the liquid.

In this work original data are reported about surfactants adsorption at superhydrophobic surfaces prepared by spin coating a glass slide with FAS-TEOS (FluoroAlkylSilane-TetraEthOxySilane) copolymer followed by thermal treatment (Fig. 2).

In air the rms roughness of this polymer was found to be 0.4 nm with a geometrical structure that can be related to a regular distribution in terms of size, shape and spacing of pillars-like structures.

Water CA in air was 147° ± 2 indicating a homogeneous surface with respect to the preparation method.

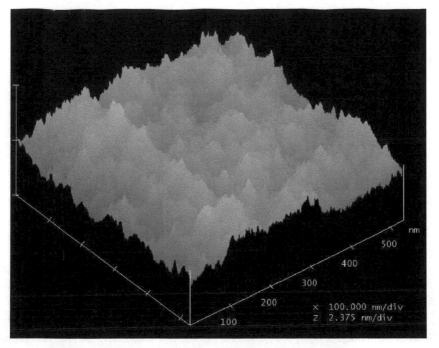

Figure 2. AFM image in air of FAS-TEOS copolymer coating on glass surface.

CA measurements were performed following the spreading kinetics in a water vapour saturated cell in order to avoid significant reduction of the drop volume during the experiment. Dynamic CAs were monitored for quite a long time (20 min) in order to allow the self-assembled structure under the drop to form a stable aggregate.

In order to study the structures of the surfactant adsorbed at this specific surface, in-liquid investigations by AFM were performed chosing a specific probing technique, minimally invasive, like soft-contact mode.

In presence of surfactants the ordered amphiphilic pattern observed with the graphite substrate is not found with this coating and the morphological study points out that the adsorption of the surfactant from the solution modifies the solid surface structure as in a smoothing process in the case of the non-ionic n-dodecyl dimethyl phosphine oxide ($C_{12}DMPO$) (Fig. 3).

Solutions of different surfactant nature such as $C_{12}DMPO$ and SDS in the presence of NaCl as electrolyte at the critical micellar concentration have been found to provide approximately the same equilibrium surface tension value, but we observe that the shape of the self-assembled structure under the drop seems to produce significant difference in CA values (around 20 degrees) (Table 1).

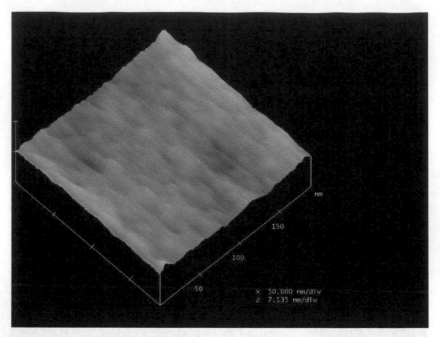

Figure 3. AFM image in liquid of C_{12}DMPO at supramicellar concentration on FAS-TEOS coated glass surface.

Table 1.

Surface tension and CA of ionic and non-ionic surfactant solutions on FAS-TEOS and superhydrophobic nanoparticles-polymer coating (SH) (graphite and water as references)

	Eq. Surface tension (mN/m) (20°C)	CA (°) on FAS-TEOS	CA (°) on SH coating
Water	72.5	147 ± 2	169 ± 1
SDS solution (c>cmc)	37.0	81 ± 1	126 ± 1
C_{12}DMPO solution (c>cmc)	33.0	60 ± 1	128 ± 1
CA of Water on Graphite 30°–80°*			

* Literature data: the wide range is due to surface contamination.

Supramicellar concentrations for both surfactant solutions have been used in order to provide the minimum value for the surface tension at the air-water interface and thus the maximum spreading efficiency on the solid substrate.

Large differences among CAs on superhydrophobic surfaces were also observed by Mohammadi et al. [36] on comparing surfactants solutions at high concentrations with different liquids having similar surface tensions.

The comparison between surfactant solutions and pure liquids in terms of wetting properties can be explained through the change in the surface energy of the solid

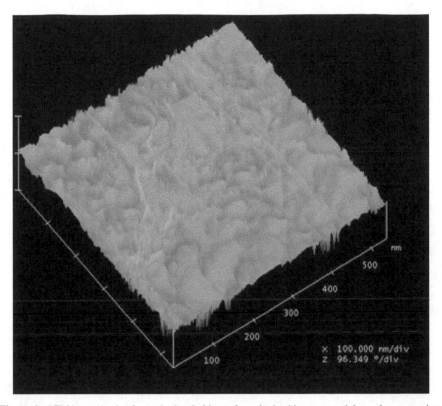

Figure 4. AFM image in air of superhydrophobic surface obtained by nanoparticles-polymer coating.

AKD (Alkylketenedimer) due to the adsorption of the surfactants on the surface, significantly influencing the CA measurements. The surfactant interaction with the AKD surface does not allow the solution to fill the surface capillary grooves by adsorbing at pore walls. As result, high CAs can be found for surfactant solutions in comparison to pure liquids with similar low surface tensions.

The significant CA variation suggests that manipulation of physical and chemical parameters is possible in order to confine the volume of the drop in a restricted space by means of a kind of "fine tuning" method. This result can be applied to control the spreading of liquids by controlling, for instance, the concentration and nature of the amphiphilic molecules.

Ferrari *et al.* [37] studied more recently the wetting behaviour of surfactant solutions on a superhydrophobic surface in order to define the role of amphiphilic molecules in modifying, by adsorption, surfaces with particular microstructure geometry (Fig. 4) such as those showing high water repellence with a potential self-cleaning application (Fig. 5).

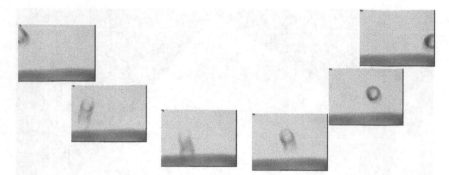

Figure 5. Picture sequence of a 4 μl water drop falling and rolling off the superhydrophobic surface prepared by nanoparticles-polymer coating.

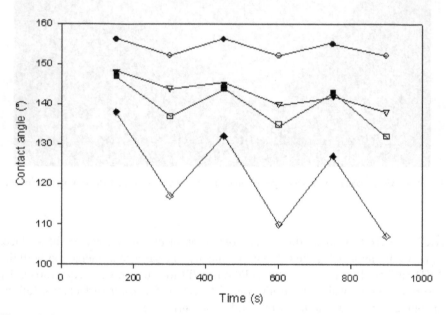

Figure 6. Dynamic CAs on mixed organic-inorganic superhydrophobic substrate of different CTAB solutions as a function of concentration and presence of salts. Advancing (filled symbols) and receding (empty symbols) CAs for CTAB aqueous solutions: $c = 2.0 \cdot 10^{-5}$ M (circles), and $c = 2.0 \cdot 10^{-5}$ M + NaCl 20 mM (triangles), $c = 1.6 \cdot 10^{-2}$ M (squares), and $c = 1.6 \cdot 10^{-2}$ M +20 mM NaCl (diamonds).

This superhydrophobic surface coating was prepared by using the original methodology described elsewhere [38] based on the deposition of a silica particle/fluorinated polymer coating giving CAs of pure water on such a surface on the order of 169°, with hysteresis <2°.

At lower (submicellar) concentrations, a superhydrophobic range was found for the surfactant solutions independent of their chemical nature, showing high CAs and a small hysteresis.

In addition, a decreasing trend of CAs with the interfacial tension was observed, pointing out the secondary role played by the adsorption at the solid-liquid interface, as the basis for the CA hysteresis, and reducing in the Cassie-Baxter approach where the most involved interface is liquid-gas.

At concentrations higher than the cmc, the electrical characteristics of the surfactants can be regarded as the origin for the high differences in CAs and hysteresis. In fact, in spite of similar values of liquid-air surface tension, the absence of salt keeps anionic and cationic surfactants still in a superhydrophobic range, while a hydrophobic behaviour with lower CAs and larger hystheresis is attained by adding salt (Fig. 6) and by non-ionic surfactant solutions (Table 1).

In this case surfactants are effective in controlling the wetting in a hydrophobic or still superhydrophobic range: a transition from a Cassie-Baxter regime to a Wenzel regime for superhydrophobic surfaces results by coupling surfactant adsorption with the strong water repellence of these surfaces, and the spreading of solutions can then be effectively controlled by confining the drop in a limited area.

4. SURFACTANTS ADSORPTION AT SUPERHYDROPHOBIC SURFACES IN LIQUID-LIQUID SYSTEMS

Liquid spreading and liquid motion control are very important in microfluidics, printing, and other applications where a solid substrate is in contact with different liquid phases and a selectivity with respect to a particular kind of liquid is required. Experimental studies regarding the absorption and desorption of organic liquids in elastic superhydrophobic silica aerogels are reported by Rao *et al.* [39]. The aerogels were prepared using methyltrimethoxysilane (MTMS) in a sol–gel process producing monolithic gels used as absorbents for organic liquids like alkanes, aromatic hydrocarbons compounds, alcohols and oils like kerosene and petrol. The uptake of organic compounds in terms of capacity and rate was very high for the superhydrophobic aerogels. The desorption of solvents and oils was studied at various temperatures until complete desorption of the absorbed liquid. From TEM observations the hydrocarbons and alcohols absorption did not affect the aerogel structure, while the absorption of oil produced a shrinkage with the formation of a dense structure after desorption. Hydrophobic properties were maintained in all cases by the gels which could be then re-used.

In the microfluidics field, for instance, small volumes of liquids flow through micro-channels toward a specific destination for further processing. The capillary properties thus play a key role in those systems where surfaces with high chemical inertness and strong liquid repellence are required, making wetting and spreading very difficult to control.

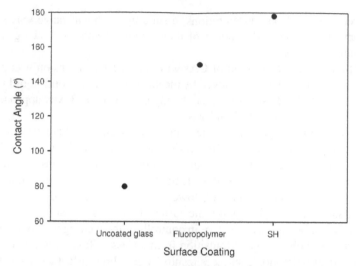

Figure 7. CA of water in hexane as a function of the surface coating: uncoated glass, fluorinated polymer, superhydrophobic nanoparticles-polymer coating (SH).

The problem of controlling droplets moving inside small capillary systems has been the subject of numerical simulations [40] conducted to characterize the effects of an insoluble surfactant on a drop in shear flow. In this work, the drop was considered as suspended in a matrix liquid and a specific parameter, related to the decrease in effective surface tension, was used to measure the surfactant amount at the interface. In a model system, it was found that adding surfactant caused stable drops to be more elongated and less inclined with respect to the primary flow direction in comparison with drops without surfactant. This behaviour was explained by the distribution of the surfactant at the interface during the evolution of the drops, which were characterized by flattened angle, shorter necks and faster time to break in comparison with similar drops without surfactant.

As usual for capillarity-related applications, specific surface active amphiphiles can be efficiently added to the liquids to tune the wetting and spreading on these interfaces. Especially in the presence of a second liquid phase, the properties of surfactant solutions on superhydrophobic solid surfaces are not yet supported by related studies and consequently by suitable reference data, providing only a partial overview in this area.

Thus new insights have come from studies of wetting properties of surfactant solutions on superhydrophobic surfaces in water-oil systems. In Fig. 7 it is shown that the increased topography, moving from the simple polymeric fluorine-based coating to a mixed inorganic–organic layer [see ref. 38], increases the CA in a water-hexane system by almost 40°, making a drop completely rolling off from the surfaces without sliding. In addition, preliminary results on the role played

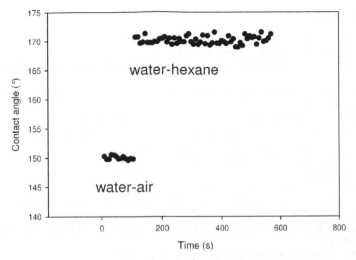

Figure 8. CAs of $C_{12}DMPO$ ($c = 2 \cdot 10^{-8}$ mol/cm^3) water solution in air and in hexane (partition coefficient Kp = 7.7) on superhydrophobic nanoparticles-polymer surface coating as a function of time.

by the partitioning of surfactants between two immiscible phases [41] such as water and hexane are presented in Fig. 8. The effect of the transfer of the $C_{12}DMPO$ to the oil phase on the CA on this superhydrophobic coating is compared with the same solution in air. These results open the avenue for the control of wetting by surfactant solutions in liquid-liquid systems: for example the switching effect from a Cassie-Baxter to a Wenzel regime obtained in air, as explained above, can be effectively reversed, bringing back the surface from hydrophobic to superhydrophobic behaviour in presence of a significant distribution of the surfactant between the liquid phases.

5. CONCLUSIONS

In this work the wetting properties of surfactants solutions in contact with solid surfaces of different degrees of hydrophobicity have been reviewed.

The role played by the composition of aqueous solutions of ionic and non-ionic amphiphiles and the geometry of the surface, in case of different surface coatings, has been described with respect to application and basic research fields.

These studies point out how surfactant adsorption and surface coating characteristics can be finely exploited in order to attain a specific behaviour of liquid-solid systems.

The original research data reported here deal, in particular, with the possibility to control wetting properties as a function of the surfactant partition coefficient in liquid-liquid systems.

REFERENCES

1. J. Bico, C. Marzolin and D. Queré, *Europhys. Lett.* **47**, 220 (1999).
2. D.Öner and T. M. McCarthy, *Langmuir* **16**, 7777 (2000).
3. L. Feng, Y. Song, J. Zhai, B. Liu, J. Xu, L. Jiang and D. Zhu, *Angew. Chem. Int. Ed.* **42**, 800 (2003).
4. L. Feng, S. Li, Y. Li, H. Li, L. Zhang, J. Zhai, Y. Song, B. Liu, L. Jiang and D. Zhu, *Adv. Mater.* **14**, 1857 (2002).
5. J. Zhang, J. Li and Y. Han, *Macromol Rapid Comm.* **25**, 1105 (2004).
6. C.-H. Wang, Y.-Y. Song, J.-W. Zhao and X.-H. Xia, *Surface Sci.* **600**(4), 38 (2006).
7. M. Taniguchi, J. Pieracci and G. Belfort, *Langmuir* **17**, 4312 (2001).
8. N. A. Patankar, *Langmuir* **20**, 8209 (2004).
9. B. He, N. A. Patankar and J. Lee, *Langmuir* **19**, 4999 (2003).
10. N. A. Patankar, *Langmuir* **19**, 1249 (2003).
11. C. J. van Oss, R. F. Giese and A. Docoslis, *J. Dispersion Sci Technol.* **26**, 585 (2005).
12. E. J. Wanless and W. A. Ducker, *J. Phys. Chem.* **100**, 3207 (1996).
13. S. Bandyopadhyay, J. C. Shelley, M. Tarek, P. B. Moore and M. L. Klein, *J. Phys. Chem. B* **102**, 6318 (1998).
14. E. J. Wanless and W. A. Ducker, *Langmuir* **13**, 1463 (1997).
15. W. A. Ducker and E. J. Wanless, *Langmuir* **12**, 5915 (1996).
16. R. A. Johnson and R. Nagarajan, *Colloids Surfaces A* **167**, 31 (2000).
17. G. G. Warr, *Curr. Opin. Colloid Interface Sci.* **5**, 88 (2000).
18. L. M. Grant and W. A. Ducker, *J. Phys. Chem. B* **101**, 5337 (1997).
19. L. M. Grant, F. Tiberg and W. A. Ducker, *J. Phys. Chem. B* **102**, 4288 (1998).
20. L. M. Grant, T. Ederth and F. Tiberg, *Langmuir* **16**, 2285 (2000).
21. M. Ferrari, F. Ravera, M. Viviani and L. Liggieri, *Colloids Surfaces A* **249**, 63 (2004).
22. H. N. Patrick and G. G. Warr, *Colloids Surfaces A* **162**, 149 (2000).
23. H. N. Patrick, G. G. Warr, S. Manne and I. A. Aksay, *Langmuir* **13**, 4349 (1997).
24. H. N. Patrick, G. G. Warr, S. Manne and I. A. Aksay, *Langmuir* **15**, 1685 (1999).
25. P.-S. Tsai, Y.-M. Yang and Y.-L. Lee, *Langmuir* **22**, 5660 (2006).
26. R. N. Wenzel, *Ind. Eng. Chem.* **28**, 988 (1936).
27. A. B. D. Cassie and S. Baxter, *Trans. Faraday Soc.* **40**, 546 (1944).
28. P. Balasuwatthi, N. Dechabumphen, C. Saiwan and J. F. Scamehorn, *J. Surfactants and Detergents* **7**(1), 31 (2004).
29. M. Payne, C. Maldarelli and A. Couzis, *Paper presented at the March 2007 American Physical Society Meeting* in Denver, CO (USA) (2007).
30. A. Puttharak, S. Chavadej, B. Kitiyanan and J. E. Scamehorn, *Paper presented at the 16th Surfactants in Solution (SIS) Meeting* in Seoul (Korea) (2006).
31. J. S. Shibuichi, T. Yamamoto, T. Onda and K. Tsujii, *J. Colloid. Interface Sci.* **208**, 287 (1998).
32. W. Barthlott and C. Neinhuis, *Planta* **202**, 1 (1997).
33. W. Ming, D. Wu, R. van Benthem and G. de With, *Nano Lett.* **5**, 11 (2005).
34. H. Y. Erbil, A. L. Demire, Y. Avci and O. Mert, *Science* **299**, 1377 (2003).
35. H. M. Shang, Y. Wang, K. Takahashi and G. Z. Cao, *J. Mater. Sci.* **40**, 3587 (2005).
36. R. Mohammadi, J. Wassink and A. Amirfazli, *Langmuir* **20**, 9657 (2004).
37. M. Ferrari, F. Ravera, S. Rao and L. Liggieri, *Appl. Phys. Lett.* **89**, 053104 (2006).
38. M. Ferrari, F. Ravera and L. Liggieri, *Appl. Phys. Lett.* **88**, 203125 (2006).
39. A. V. Rao, N. D. Hegde and H. Hirashima, *J. Colloid. Interface Sci.* **305**, 124 (2007).
40. M. A. Drumright-Clarke and Y. Renardy, *Phys. Fluids.* **16**(1), 14 (2004).
41. F. Ravera, M. Ferrari, L. Liggieri, R. Miller and A. Passerone, *Langmuir* **13**, 4817 (1997).